Integrating Gender into Transport Planning

"Books to provide transport planners with gendered insights of everyday mobility are rare. It is an important but under-researched topic because although we know that satisfying people's mobility needs is about much more than improving journey efficiency transport planning decisions are still largely focused on this issue. The authors of this book demonstrate that by applying feminist approaches to the study of mobility there is a chance to open new windows of understanding about how and why people are mobile or immobile and whether their daily experiences of the transport system matches their needs and expectation. Women, in particular, but anyone who adopts the 'feminine' role within society will have different travel demands to those traditionally catered for by transport professionals. The authors of this book recommend that it is time to establish new planning regimes that recognise the diversities that are manifest within contemporary societies in order to move beyond the binary categorisations of commuter and social travel and towards multivariate travel environments that recognise the needs of women, but also children, older people, ethnic minorities and different economic stratifications. The book is well-placed to offer useful new research approaches that will assist travel planners in this task."

—Karen Lucas, *Deputy Director of the Leeds Social Sciences Institute*
and Professor of Transport and Social Analysis,
University of Leeds, UK

Christina Lindkvist Scholten
Tanja Joelsson
Editors

Integrating Gender into Transport Planning

From One to Many Tracks

Editors
Christina Lindkvist Scholten
Malmö University
Malmö, Sweden

K2 – Swedish Knowledge Centre for
Public Transport, Lund, Sweden

Tanja Joelsson
Department of Education
Uppsala University
Uppsala, Sweden

ISBN 978-3-030-05041-2   ISBN 978-3-030-05042-9   (eBook)
https://doi.org/10.1007/978-3-030-05042-9

Library of Congress Control Number: 2018964070

This Palgrave Macmillan imprint is published by the registered company Springer Nature
Switzerland AG
The registered company address is: Gewerbestrasse 11, 6330 Cham, Switzerland

ACKNOWLEDGEMENTS

The work behind this collection has been a collective process and has been characterized by passion for gender research, equity and the curiosity and ambition to bring different strands of research traditions together. Discussions on gender, transport and planning have also initiated discussions on power relations and taken-for-given norms regarding transport planning. The everyday experiences of transports and mobility and how planning structures people's everyday life are a part of these discussions. Also, how to integrate qualitative and quantitative research traditions to emphasize the importance of different perspectives to contribute to solid knowledge on gender, transport planning and mobility has been an explicit ambition.

The collection would not have been possible without the support from different actors: the management team at K2—the Swedish Knowledge Centre for Public Transport, which supported the original idea and financially contributed to bring most of the authors together and, by doing this, developed an important network for future research; the WSP Group in Stockholm, which hosted some of our meetings; friends and colleagues who have been encouraging and helpful during the process, by reading and commenting on draft manuscripts. All the contributions in the collection are original research financed by national research grants and funding in Sweden, Denmark, Spain and the United Kingdom, and to which we are grateful. We also would like to thank the journals *Urban Studies* and *Environment and Urbanization* for the permission to republish the articles by Professor Caren Levy and Dr. Erika Sandow.

The publishing team at Palgrave Macmillan has made the project possible by connecting us with Ms. Rachel Daniel and Ms. Kyra Saniewski, who identified the potentials in the first draft. The comments from the anonymous reviewer of the first draft manuscript have been of immense help in developing the theoretical framework of the collection.

The language editing by Dr. Liz Sourbut has improved the quality of the contributions in every way.

Finally, we thank the contributors for their patience and for never giving in during the process.

Christina Lindkvist Scholten
Tanja Joelsson

CONTENTS

NOTES ON CONTRIBUTORS

Dag Balkmar PhD, is a researcher, Centre for Feminist Social Studies, School of Humanities, Education and Social Sciences, Örebro University, Sweden. Balkmar has a PhD in Gender Studies (Critical Studies of Men and Masculinities) and is conducting research on intersectionality, mobility, policy and violence at the Centre for Feminist Social Studies, Örebro University, Sweden. His research interests concern gender, mobility and risk, including cyclists' experiences of their everyday environment in relation to mobility, violence and politics.

Hilda Rømer Christensen PhD, is an associate professor and head of the Co-ordination for Gender Research at the University of Copenhagen, Denmark. She has written extensively on gender, culture, religion, welfare and citizenship in historical and current perspectives and is the editor-in--chief of the research journal *Women, Gender and Research*. Christensen has been the scientific coordinator for the FP6-funded SSA TRANSGEN focused on gender mainstreaming of transport and research in the EU. Since 2010, she has been a member of the Co-ordination committee for Welfare and Innovation at the Sino-Danish University Centre in Beijing and heads a research project entitled 'World Dynamics in Micro Perspectives: (Re)making Middle Class Families, China—Denmark, Including Housing, Mobility and Transport'.

Inés Sánchez de Madariaga Arch, PhD, MSc, is Professor of Urban Planning at Madrid School of Architecture, Spain, and Director of the UNESCO Chair on Gender. She is the founder and director of the first Spanish research group into gender, urban planning and architecture. She has been PI and/or member of the scientific committee of over 25 research projects addressing gender dimensions in the following areas: planning, development, STEM, architecture and engineering. She is a member of the Board of Advisors of the European Women Rectors Association, Chair of the Research and Academia Partner Constituency Group of the General Assembly of Partners of the New Urban Agenda of UN-Habitat, and a member of the Leadership Council of the Spanish Chapter of the UN-Sustainable Solutions Development Network. She has authored over 60 scientific articles and book chapters and 6 books, and lectures around the world.

Charlotta Faith-Ell holds a PhD in Land and Water Resources Engineering from the Royal Institute of Technology and a MSc in Earth Sciences from Stockholm University. Faith-Ell has more than twenty-five years of experience of working with project management and environmental management. Faith-Ell is a recognised expert in the field of Impact Assessment. She is Senior Technical Director of Environmental Impact Assessment (EIA) and Strategic Environmental Assessment (SEA) and Director of Research at WSP Civils in Sweden. She also holds a position as researcher at the Department of Ecotechnology and Sustainable Building Engineering the Mid-Sweden University.

Clara Greed PhD, is emerita professor of Urban Planning, at FET—Architecture and the Built Environment, University of the West of England, the United Kingdom. She is a member of the Royal Town Planning Institute and other built environment bodies. Greed's research activities are focused upon the built environment, architecture, planning and urban social issues. Among other things, her research has concerned public toilet provision within the context of urban design and the social aspects of planning. In recent years she has returned to her previous interest in urban theology and the relationship between religion and urban structure, and the emergence of the post-secular city.

Malin Henriksson PhD, is a researcher at the Swedish National Road and Transport Research Institute (VTI), Linköping, Sweden. Henriksson holds a PhD from Linköping University within the area of technology and social change. Her thesis from 2014 develops an intersectional framework on planner's images of sustainable mobility and combines qualitative methods with feminist planning theory. Henriksson's research concerns transport, mobility and power.

Tanja Joelsson holds a PhD in Gender Studies from Linköping University, Sweden, and is appointed as a senior lecturer and researcher in Child and Youth Studies at the Department of Education, Uppsala University, Sweden. Joelsson's research concerns child and youth cultures, sociocultural studies on risk and affect, and children's and families' everyday mobility predominantly from ethnographic vantage points.

Lena Levin PhD, is a researcher at the VTI, Linköping, and K2, Sweden. Levin is a researcher at the research unit Mobility, Actors and Planning processes (MAP) within the VTI, and at K2, the Swedish Knowledge Centre for Public Transport, Sweden. Her research centres on how the transport system is shaped, developed and utilised by actors with various interests, perspectives and ascendancy. Levin has conducted research on gender mainstreaming (e.g. Gender Impact Assessment, GIA) and social issues (e.g. Social Impact Assessment, SIA) in transport infrastructure planning.

Caren Levy PhD, is Professor of Transformative Urban Planning, at the Development Planning Unit, University College London, the United Kingdom. Levy's research is focused on interdisciplinary and applied knowledge production in the context of development theory, policy and planning practice in a range of organizational landscapes, both institutional and geographical. The common theme in all is that of methodology development, focusing on the institutionalization of social justice in policy and planning in the context of inequalities in the global South.

Erika Sandow PhD, is a researcher at the Centre for Demography and Ageing Research (CEDAR) and Senior Lecturer in Spatial Planning at the Department of Geography and Economic History at Umeå University, Sweden. Her research has mainly focused on how social and geographical conditions cause different prerequisites and needs for long-distance commuting and social consequences related to long-distance commuting in terms of gender differences in commuting patterns and earnings, separations and health consequences of commuting. Sandow's research focuses on the relationship between the geography of family networks, social mobility and migration.

Christina Lindkvist Scholten PhD, is a senior lecturer and researcher and Associate Professor of Urban studies, Malmö University, and K2, Sweden. Scholten holds a PhD in human and economic geography from Lund University, Sweden. Most of Scholten's research focuses on gender equality, gender politics and planning. Scholten is involved in several projects on transport, mobility, urban planning, collaboration and gender equality.

Lena Smidfelt Rosqvist is Research Director at Trivector Traffic AB, Lund, Sweden. Smidfelt Rosqvist has a PhD from Lund University, Sweden, and her expertise is on sustainable transport behaviour and transport systems (environmental effects, traffic assessments, gender equality and travel surveys and impact calculations for various development scenarios). Smidfelt Rosqvist was the programme director of the interdisciplinary research programme *TransportMistra*, aimed at creating sustainable transport systems. She has also been Chairman of the non-political and nonreligious organization *Nätverket för kvinnor i transportpolitiken* (Network for Women in Transport Politics) aiming for achieving the objective in Swedish national policy on gender equality in the transport sector (2010–2015).

Elena Zucchini is a town planner, with an MSc and PhD in Town Planning at the University of Madrid with thesis "Gender and transportation: analysis of mobility of care as a starting point to build a knowledge base of mobility patterns. The case of Madrid". Zucchini has also been a visitor researcher at the Institute for Transport Studies in Leeds, United Kingdom, and is a member of Bristol Women's Voice and of the Architecture Centre of Bristol. She has also more than ten years of experience as a town planner in a private consultancy and has developed medium- and large-scale projects around the world.

List of Figures

LIST OF TABLES

LIST OF BOXES

The Political in Transport and Mobility: Towards a Feminist Analysis of Everyday Mobility and Transport Planning

Tanja Joelsson and Christina Lindkvist Scholten

INTRODUCTION

Transportation permeates our daily lives and produces and organizes social life. Transport politics and transport planning also constitute fundamental cornerstones in the building of modern society. Accordingly, transport planning is inherently political. For a long time, transport policy and planning have been generally understood as a matter of technical practice designed to find solutions for how to bring people, places and goods closer to one another and reducing the cost of doing so. From a Euclidian perspective, physical distance becomes measurable (Healey 2006), and an ordinary everyday trip can be split into sub-journeys dependent on length,

T. Joelsson (✉)
Department of Education, Uppsala University, Uppsala, Sweden
e-mail: tanja.joelsson@edu.uu.se

C. L. Scholten
Malmö University, Malmö, Sweden

K2 – Swedish Knowledge Centre for Public Transport, Lund, Sweden
e-mail: christina.scholten@mau.se; christina.scholten@k2centrum.se

© The Author(s) 2019
C. L. Scholten, T. Joelsson (eds.), *Integrating Gender into Transport Planning*, https://doi.org/10.1007/978-3-030-05042-9_1

1

motive for the trip, mode of transport, travel time and whether the traveller is with someone else or not. However, mobility is just as much about the expectations, experiences, emotions and meanings that people attach to movement in various situated contexts and situations. Feminist researchers have opposed the mainstream transport planning paradigm (see Greed in this collection, Matrix 1984; Oliver 1988) by pointing out that mobility is always situated. The particularities of mobility imply that those engaged in movement cannot be reduced to flows or numbers, but must always be considered as embodied and material. Men, women, boys and girls travel, move and use the transport system on an everyday basis, and are both affected by and affect this movement and the surrounding environment. Awareness of the situated character of mobility is demanding and challenging for both decision-makers and planners. Governments have also started to acknowledge the importance of considering gender in transport policy-making (see Swedish transport policy objectives and Levin and Faith-Ell in this collection). However, even in Sweden, a country often described as progressive in implementing gender equality, much needs to be done to improve gender-equal transport planning. A possible explanation as to why transport planning still lags behind in integrating gender into its planning practice is the utilitarian planning tradition in which the emphasis is on organizing and structuring transport for the good of businesses and the public. In doing so, the rationale for decision-making about investments has been based on a tradition of reductionist knowledge; moreover, the transport system has been implemented via top-down planning practices. To be able to calculate the net effects of investments in transport, complex physical and social relations have been reduced to measurable numbers and calculable flows of goods and people. As a consequence of the elaborate calculations and transport modelling carried out by transport planning experts, ordinary people face difficulties in reviewing or assessing the infrastructure investments, or valuing the outcomes of such projects.

With growing concern about the environment and climate change, transport planning as we know needs to develop and incorporate knowledge perspectives which until now have been largely absent. Global urbanization draws attention to more integrated transport and urban land-use planning, in which perspectives grounded in qualitative research traditions can help transport planning to address problems which are not easily reduced to isolated and measurable numbers. Transport planning needs to take into account *both* the numbers that enable us to calculate, make prognoses and measure flows *and* user experiences and expectations and the meanings of the transport system. This edited collection is a contribution to a more holistic transport planning practice aimed at bringing pluralistic planning traditions and perspectives together. The collection contributes to the development of a transport planning approach that better understands the qualities and meanings of the transport system to a multiplicity of users. This will hopefully aid in transforming transport policy-making and transport planning, and eventually the transport system itself, in more inclusive, equal and efficient directions.

USING GENDER AS AN ANALYTICAL TOOL IN TRANSPORT PLANNING

The concept of gender, as Hanson describes, has "a long and complex genealogy" (Hanson 2010: 8). Lykke (2009) discusses the birth of Euro-American gender research as being related in part to gender as a topic of study, and in part to various degrees of academic institutionalization of the field. In this process, she claims, there has always been a dual agenda of developing the research field, and remaining cautious as to how institutionalization may affect the field. Gender research is hence a *post-disciplinary field*, researching topics that need transdisciplinary and interdisciplinary approaches stretching beyond disciplinary boundaries (Lykke 2009). Transport and mobility constitute just such a hybrid phenomenon (Latour 1993), we argue, which needs to draw upon theoretical frameworks and tools from several conventional disciplines as well as from post-disciplinary fields, in order to produce novel understandings of complex realities.

In a similar vein, using gender as an analytical tool in relation to transport and mobility necessitates a balancing act between different theoretical feminist ways of using the concept of gender. Gender can also be regarded as a hybrid concept in line with Latour's understanding (Lykke and Braidotti 1996), which can be illustrated by the scholarly sex/gender division in Anglo-American theorizing and related to further theoretical developments of gender as a concept (Lykke 2009). What is relevant to this collection is, firstly, that feminist analyses of gender are and have been carried out in contrast to gender-conservative ideas and conceptions of gender, thus critiquing and deconstructing determinist and causal connections between biological sex and the social construction of gender. Secondly, perceiving gender as socially constructed explicitly foregrounds the situatedness and contextual character of gender positions, organized in line with societal and cultural norms on accepted performativity (Butler 2006; Rubin 1975). Gender in this sense can never be fixed or stable, but is always open and vulnerable to 'performative failures' whereby it can be undone or re-done. The "doing" of gender (West and Zimmerman 1987) has resulted in elaborate insights into the sociocultural understanding of gendered positions. Thirdly, gender as an analytical tool, and as the object of study in feminist research, is continuously under scholarly scrutiny within the gender research field. The genealogy of the concept is, however, important to bear in mind when we move on to the different chapters of this book.

The cultural and sociopolitical understanding of gender and power relations needs to be connected to the practices of planning, in which respect, integration and acceptance of diversity are cornerstones. In planning, it is crucial to measure and to calculate. For this purpose, numbers are needed and categorizations need to be made whereby one category is strictly separated from another. Two social categories frequently used in transport planning are 'women' and 'men', which in light of the discussion above can be critiqued as fixing people and identities and simplifying complex realities. Given the precedence that quantitative studies and travel surveys are given in transport planning, gender research based on post-structural and post-modern epistemologies studying more qualitative aspects of everyday life may encounter difficulties in the translation process. Matters are further complicated by the tensions and challenges encompassing gender equality as a political objective (understood in relation to statistical data on men and women) and critical feminist research, although the goal of moving towards an accessible transport system is the same. We argue that both qualitative and quantitative perspectives are needed in order to develop transport politics and a transport system that take account of people's different needs. And it is necessary to enhance the dialogue among researchers working from different epistemological standpoints. With the ambition of contributing to a better-informed transport planning and policy-making, the starting point for this collection is people's everyday lives and mobility. Hanson formulates this ambition in a clear way:

> I want to stress that mobility is not just about the individual (...), but about the individual as embedded in, and interacting with, the household, family, community and larger society. That is, it should be impossible to think about mobility without simultaneously considering social, cultural and geographical context—the specifics of place, time and people. (Hanson 2010: 8)

Planning of the built environment and the transport system has histori-cally been a man's world (see Greed in this collection). Even though this is changing, with more women entering the transport planning profession, the norms, standards and institutional conditions under which transport planning is conducted remain fairly solid (Greed and Levy in this collec-tion). The transportation of individuals and their dependants, often children and minors, to everyday destinations is structured and managed by the poli-tics that shape the spatiality of the transport system. Their experiences will be based on the interests and needs of those who have determined the start-ing points for designing the system. Moreover, the transport system will be experienced differently according to the resources and subject position of the individual traveller. Feminist-informed transport planning needs to be rooted in an understanding of the following: the efforts behind citizens' desires and needs to overcome distances that are built into the spatial struc-tures of society; and how the spatiality of the transport system supports or disconnects citizens from appropriating public space and access to services and amenities in order to live life independently. To understand these efforts, transport planners must have a substantial understanding not only of gen-dered power relations but also of the power relations that are created and established through a variety of subject positions, usually referred to as intersections. Moreover, transport planners need an understanding of how these power relations structure one's ability to use the transport system.

Since the 1990s, new perspectives on transport have developed. In par-ticular, research based on critical and feminist epistemologies has revealed important perspectives on transportation as indeed being a gendered prac-tice. The gendering of transport planning and transport itself encompasses everything from travel mode choices to the purposes of the trips made by women and men. This research has visualized the gendered conditions of everyday life (Hanson and Pratt 1995; Law 1999; Dobbs 2007; Schwanen et al. 2008; Uteng and Cresswell 2008; Hanson 2010; Balkmar 2012; Scholten et al. 2012; Joelsson 2013). By looking beyond the statistical and disembodied traveller, feminist research has identified the preconditions for travel for different groups, for example, the elderly (Levin 2008; Levin and Faith-Ell 2011; Berg et al. 2014) or young people (Brown et al. 2008; Barker et al. 2009; Salon and Gulyani 2010; Fyhri et al. 2011). By starting with an understanding of society as built on power relations and gendered structures, feminist research and epistemologies have evolved into a mani-fold of perspectives and analytical conceptualizations of gender (Hemmings 2005; Butler 2006).

In this collection, the different contributions represent the variety of ways in which the concept of gender is used to analyse the outcomes of transport planning. In the chapters using travel surveys and gender mainstreaming, statistics based on sex categories are necessary to develop knowledge about what conditions women and men are facing in relation to travel habits and the outcome of transport planning decision-making. Clearly, there are fundamental organizing principles connected to power that need to be identified: for example, how important everyday destinations are made accessible, to whom and by what mode of transport; or which infrastructure investments favour which kinds of travel group; or who is actually the main person in a household responsible for escorting minors or the elderly to their regular activities; or who is the long-distance commuter and who is not. Interpreting and analysing the role of gender in these phenomena needs a gender-sensitive approach within which sociocultural norms and values are related to the practices of real people and how constructions and conceptions of gender position men and women differently and unequally.

Other contributions in this collection elaborate upon how gender becomes constructed through transport planning practices. In these chapters, there is no clearly identified gendered position; instead, gender constitutes itself through different actions and situations in which normativity, performativity and context offer a gendered identity position to be adopted or claimed. In these examples, individuals are carving out a space in which they can develop strategies on how to act, react and interact with others in the transport system, such as consumers of transport services, as users of the transport system or its planners. Advertising and consumerism as discursive constructions of the gendered transport user are important structures in the process of trying to establish or nail down specific gendered positions within the transport system, which transport users try to avoid. The analytical outcomes of the social construction of transport system users are important contributions to bringing transport planning practice into diverse and complex settings where issues relating to the embodied traveller, norms concerning the design and development of transport planning and the processes defining what problems transport planning is supposed to solve. The somewhat contradictory, and to some extent less coherent, way of addressing and understanding the concept of gender in this collection is thus a pragmatic solution to acknowledging the structuralist as well as post-structuralist research agendas of gender and transport planning, which is in line with the ambition to integrate rational research traditions with a social constructivist approach. By providing an opportunity to introduce texts in which gender and a gender perspective are used in their broadest understanding, our ambition is to enhance and encourage as many ways as possible of dealing with the importance of gender in transport planning.

Several of the articles in this collection are inspired by the works of Chantal Mouffe (2005, 2013) and Carol Lee Bacchi (2009), through addressing the political in designing the transport system, allocating resources, conducting priority-making and constructing the user of the transport system. It is, however, important to also include the embodied planner of the transport system as a person with expert skills, experiences, norms and values regarding what constitutes the planning profession and also what is considered to be good planning and desirable outcomes of the planning process. Despite the various and sometimes contradictory claims, the overall message is that transport politics, policy-making and planning are essential to people. Different political interests have to be acknowledged. Furthermore, models on how to integrate participation from a broader stakeholder representation in the political process of transport planning and policy-making need to be developed to ensure that user groups who are seldom involved in the planning process are listened to and encouraged to participate (cf. Listerborn 2007).

The utilitarian and consensus-driven approach silences opponents and discourages tensions among different interests (Mouffe 2013). Moreover, it can facilitate the development of instrumental outcomes of politics as a result of what Mouffe (2013) terms the "aggregative" paradigm in liberal politics, in which market-oriented economics are used as core values to understand the rationality of individuals. The point we would like to make, in line with Mouffe (2013), is that there is no generalized citizen or user of the transport system. There are different social groups and interests, and the plurality of these social groups also contains conflicting interests regarding what should be prioritized (Legacy 2016; Mouffe 2013).

To contribute to a more informed, nuanced and, hopefully, just outcome of transport politics, this collection discusses different aspects of the political in terms of shifting focus and acknowledging different user groups and standpoints taken. In the Swedish context, the decision by the government in 1994 to implement gender mainstreaming as the model to promote gender equality demands that transport planners and politicians look beyond equal numbers of men and women on decision-making bodies and also produce gender equality outcomes. Transport planners face two major difficulties when trying to translate the objectives of gender equality policy into planning: firstly, the objectives are based on the tools used in transport planning. Secondly, they struggle with how to transform the gender equality objectives into input-data in the models used. Another difficulty is how to conduct an analysis of the outcome of the planning process from a power perspective in which users of the transport system are transformed into embodied children, the elderly or cyclists. Even the planners themselves turn into embodied, gendered individuals with knowledge and experiences that impact upon the outcome of the planning process.

TRANSPORT POLITICS, POWER AND GENDER?

Fainstein and Servon (2005) speak of planners and policy-makers as public servants, whose task it is to serve the public interest. Public interest, they continue, is, nevertheless, seldom unified or single, but multiple: "Interests are multiple, as are publics" (Fainstein and Servon 2005: 2). Rosenbloom explains further:

> Planning may be thought of as a technical process in which sophisticated tools and methods are brought to bear on public problems. (…) Yet it is difficult to analyse the planning process, in the transportation or any other field, without recognizing that planning decisions involve choices between competing interests and values. Even if planning methods and tools were value-free (…), it is clear that choices between interests in conflict raise equity issues. *At the very least planners must explain to decision makers what the interests in conflict are and determine upon which groups in society the burden of a given decision will fall.* (Rosenbloom 2005: 249, emphasis added)

Organizing the transport system to respond equally to the needs of men, women and different social groups in society is a political responsibility that is executed by transport planners. As Rosenbloom (2005: 249) puts it, "Ultimately the choice between competing interests is a political one, but the proper presentation of those interests is a planning concern". In cities—characterized by urbanization and densification, complex social structures, functionalist planning heritages and growing political demands for sustainability—the development, functioning and effects of the transport systems are growing in importance (Lucas 2004, 2012). Consequently, the planning profession is further constrained and scrutinized. From empirical research, we can conclude that there is a growing interest among transport planners to learn more about how to include social dimensions and to ask new questions about the outcomes of the planning praxis (Levin et al. 2016).

Bringing a feminist perspective to the field of transport is about challenging gender-blind politics in a policy area that impacts upon the everyday lives of women and men, girls and boys. As feminist theory stretches from Marxist to post-modern analyses, it provides a broad spectrum of how to formulate and address research questions about power and subject positions. The feminist project is, therefore, a liberating social project, mainly concerned with analysing, deconstructing, challenging and reformulating power structures. Gender is a relational and dynamic concept, sensitive to both stable and changing social, spatial and cultural processes. More importantly, in our understanding, gender as an analytical category should not be understood as only dealing with gender in its narrowest layperson sense. As presented above, the social fabric of any given society is always a network of relations, and the analyses of social life must, therefore, acknowledge and be concerned with several analytical social categories and their interconnections.

When analysing gender, cultural conceptions of femininities and masculinities are central to understanding how such concepts as mobility are read and understood. According to Wolff (1993), the metaphorical use of concepts of travel in contemporary social theory has had the effect of foregrounding masculine experience. Wolff argues that we must remain cautious about the concepts we make use of and not become carried away by the seductions of androcentric metaphors that "impl[y] a notion of universal and equal mobility" (Wolff 1993: 235). As a result, the feminist critique of studies of mobility has given us tools to deconstruct how gendered assumptions and conceptions permeate research on mobility, and warns us against re-creating androcentric, ethnocentric or age-blind understandings.

Examining transport policy and planning as an institution (Martens 2016), it is notable that transport has traditionally been connected to masculinity and technology. Feminist research in science and technology identified early on how gender and technology were co-constructing each another (see e.g. Cockburn and Omrod 1993), but also highlighted the importance of more contextualized analyses for a better understanding of how technology has been described, used and developed over time (Wajcman 2010).

Furthermore, feminist transport researchers have criticized the field of transport research and practice for assuming men's travel behaviour is the norm and for neglecting women's everyday travel behaviour and patterns (Law 1999). Although some topics have been successful in addressing this male bias—most notably, research concerned with journey-to-work travel and the effect of the 'geography of fear' on women's movement—other topics have been considered less significant or not investigated at all (Law 1999). Levy (2013b: 47), for instance, has recently argued that "mainstream transport planning still remains largely untouched by debates on diversity and difference in cities". She goes on to show how issues relating to gender are still identified as a social dimension, which is marginalized from the mainstream focus on economics. In addition to the critique laid out by her feminist predecessors, she finds that:

> Overlaid on these gender-based assumptions are other transport biases, namely a focus on the (male) journey to work and other motorized transport, particularly on the private car, which is largely unaffordable to most poor urban women and men. (Levy, 2013b: 49)

This initial gender-focused critique in relation to transport has been extended to wider theoretical debates about land-use structures, the spatial separation of production and reproduction, and the binary of the public and private (Law 1999). Feminist geographers have vividly discussed and problematized spatial politics on different scales, such as spatial rights and the discourse around the right to the city (which could be rephrased as the right to public space in general). Inspired by the cultural turn in social theory—which then formed into the field of research on mobility as we see it—Law (1999) proposes that studies of daily mobility should be part of the larger project on social and cultural geographies of mobility, and that gender as a social and analytical category needs to be systematically adopted and developed. There are, then, several challenges to be met at different levels in order to advance research on both transportation and mobility, in general, and on transport planning, and the practice of transport planning, in particular. The contributions in this collection show, in different ways, how gender-sensitive approaches make a point of reflecting upon and affecting the conceptual use of gender within the domain of mobility, and how this, in turn, generates consequences for policies and politics within the transport realm.

As should be clear from the overview above, transport and mobility are imbued with politics. We perceive mobility and mobility practices as political acts rather than instrumental or technical exercises (Levy 2013a, b; cf. Law 1999). With this book, we want to reconfigure research on transport and mobility, arguing that it is crucial to reframe this subject as highly political in order to address questions of social equality, sustainability and democracy. A central predicament in relation to transport and mobility politics is to deconstruct what is perceived to be 'political' or 'politics'. Broadening the notion of what politics is and how it is done opens up opportunities to discuss political practices as both discursive and material. Bacchi's (2009) key work on policy analysis is a case in point: problematizations, in relation to policy, define and shape what we see as 'social problems'. The "issues that are problematized – how they are thought of as 'problems' – are central to governing processes" (Bacchi 2009: xi). In line with this, our intent is to identify and engage with the act of problematization and how 'problems' are represented. As Bacchi notes:

> We need to start thinking about the understanding that underpins identified problem representations. What is assumed? What is taken-for-granted? What is not questioned? (Bacchi 2009: 5)

The interest lies in how discourses about problematizations are produced—how certain phenomena are discussed and addressed and, equivalently, what is left out of the discussion. Discourses shape the ways in which phenomena are talked about, and they construct a framework for what can be expressed and how subjects (and objects) are constituted.

Planning is an act of reduction, in which political visions are narrowed down to general structures that are detailed enough to catch the idea of the vision but not specific enough to crystallize it. There is a need to be able to adjust and rework the visions. The problem, which the texts in this collection try to deal with through different perspectives, arises when the politics and practice become decoupled in such a way that transport planning fails to deliver a manifest outcome of the political visions of a gender-equal transport system. In a Swedish context, this is a problem, since the Swedish government has proclaimed a gender-equal transport system to serve the needs of both men and women equally. This is a direct outcome of the gender equality politics the Swedish government has prioritized since the mid-1980s. The gap between the political rhetoric and the transport planning outcome reveals the practical difficulties involved in producing the sustainable and gender-aware transport planning that is addressed in this collection.

FEMINIST INTERVENTIONS IN TRANSPORT PLANNING

The book is divided into four parts, ranging from chapters delving into conceptual and theoretical inquiries to contributions which are more empirical or concerned with practical tools for change. In the first part of the collection, 'Feminist Interventions in Transport Planning', two renowned scholars discuss gender and transport from slightly different vantage points. With a lifelong expertise in investigating unjust gender relations in public space and planning, Clara Greed introduces the foundations upon which many of this collection's contributions build. Greed unites the threads of feminist research in planning and transport practice, drawing the conclusion that there still remains much to be done concerning gender and transport, despite feminist research in transportation dating back to the early 1980s, which has developed the field and brought important dimensions into the research agenda of transport planning. There has been hesitation about using numbers and statistics in feminist research, mainly due to the feminist epistemological and ontological critiques of the disembodied knowledge producer and the idea that reality should be grasped and understood as something objective and manifest, seeing it instead as something produced through power relations and subject positions. This has been a necessary and important development within the social sciences and humanities. However, Greed highlights that, by examining the very methods and perspectives used in conventional and neoclassic transport planning, feminists might actually beat the masculine planning practices on 'their own turf' and challenge not only a taken-for-granted and rationalistic methodology, in general, but also the very construction of the 'objective' and 'neutral' categories and parameters used in transport planning.

Nevertheless, the feminist research agenda on transport planning needs to continue. Greed, like others in the collection, identifies that the 'social' in the very imprecise concept of 'sustainability' needs to be investigated more seriously. As Greed puts it, "aren't all policy issues social?" The necessity of working towards a gender-aware and intersectionally informed policy-making—including everyday life with all its complexity—has to become a primary matter in transport politics, as Tanja Joelsson suggests. Acknowledging the different needs and uses of public space, and the accessibility of that space, is crucial in fostering democratic inclusion. The use of the transport system as a public space is a political act. Creating safe transport spaces for cyclists and pedestrians of all ages and capabilities is a political act. Investing in roads and streets to enhance car use is also a political act. How the space for transportation is designed and equipped to match the needs of different traveller groups in different stages of life and with different capabilities is yet another political act. In her chapter, Greed presents examples from the micro-planning level, where the traffic-calming design of a street is important for the more vulnerable transport system users, and then continues on to the national level, where decisions are made about where to focus investment. A feminist and intersectional analysis is needed at all institutional levels.

Caren Levy's contribution takes an international perspective. In her text, she deconstructs the rational tradition in transport planning of travellers' choice that perceives travellers as rational individuals striving to maximize efficiency. Most people know that travelling on an everyday basis is seldom about making rational and cost-efficient decisions. Rather, it concerns using the most readily available modes of transport, if there are any. As a feminist scholar, and an early contributor to the field of gender and transport research, Levy concludes that, even after a quarter of a century, gender is still treated as 'the social', which in transport planning literature and research is considered a marginalized (and less significant) aspect compared to the field of economics. Based on empirical research in the Global South, Levy investigates the gender dimensions of transportation by critically examining spatial concepts and the distribution of amenities. By entering into dialogue with the conventional—or what Kębłowski and Bassens (2018) refer to as the neoclassical—planning paradigm, Levy rejects the idea of the disembodied traveller making rational choices between the options available. Instead, she argues that decisions about how to travel, or whether to travel at all, are part of a complex process based on available resources, vulnerability and accessibility.

INSTRUMENTS FOR CHANGE AND GENDERING TRAVEL SURVEYS

In the second and third parts of this collection, contributors engage with what could be termed potential ways forward for promoting social change in the field of transport and gender. Based on feminist and critical epistemologies, these contributions question the dominant strands of politics in the transport planning research field. This is done in order to explore the tensions that become evident when the taken-for-granted and normative prerequisites regarding transport and mobility are analysed. The aim is also to engage in critical discourse regarding whose interests are met in political transport decision-making and whose are excluded. Historically, women's transport needs and complex travel patterns have been neglected by planners designing the transport system (see Greed in this collection). The sustainability shift (Banister 2008; May et al. 2008; Hull 2008) has put forward public transport, walking and cycling as acceptable and sometimes desirable modes of transport in the transport planning of the Global North. At the same time, these forms of transport are what the poor have been using, and continue to use, compared to the better-off in both the Global North and Global South (see Levy's contribution in this collection). The major impact that more sustainable transportation behaviour would have is underestimated, according to Lena Smidfelt Rosqvist in this collection. By analysing statistics, Smidfelt Rosqvist provides evidence-based outcomes revealing that, if men began to travel by public transport and use sustainable modes of transport in comparable numbers to women, there would be net effects on the emissions from traffic at a national level.

Moreover, transport planners are concerned about responding to the national policy on gender equality as well as working towards the objectives of a sustainable urban development, within which transport planning is crucial. In terms of gender equality, since 1995, the Swedish government has adopted the method of gender mainstreaming to promote the implementation of gender equality policy in public organizations and public authorities. In 2001, gender equality was added to the national transport objectives. In 2009, the transport policy was revised, and two overall objectives were formulated into functionality and impact objectives. Gender equality is one of several aspects formulated within the objective of the functionality of the transport system. Despite Sweden's international reputation for being progressive on gender equality issues, there are difficulties regarding how to adapt and transform general gender equality objectives into planning praxis. During their ten years spent looking into the praxis of environmental assessment models, Lena Levin and Charlotta Faith-Ell have struggled with how to develop a methodology that takes transport planners at local and regional planning levels beyond checklists and catchy advertisements. They propose to achieve this by working together with transport planners in interdisciplinary research projects as a strategic way of transferring knowledge about how to integrate gender equality into the everyday praxis of transport planning. The result of this work is the Gender Impact Assessment Model, which is further introduced in the chapter by Levin and Faith-Ell. Here, the authors not only provide an overview of the nationally and internationally agreed objectives regarding gender equality as they are formulated in ratified documents by local governments and national states, but also suggest how the Swedish gender equality objectives should be operationalized to become helpful in transport planning.

Both national and global neoliberal politics also impact upon local transport planning in relation to growth-oriented and neoliberal politics in regional development. Transportation and the development of fast regional commuting have become important tools in the creation of the 'attractive' region. Regional development politics might have similar objectives, at least within the European Union, which is the supra-national level in accordance with which Sweden and other member states must act. The former national policy on regional development compensation, whereby less-populated areas of Sweden were the net receivers of government financial support, has been replaced by policies on growth. This policy emphasizes competition as a way of gathering local and regional forces. Moreover, it has become a question of local and regional public bodies, together with local industry, uniting in the rat race to attract what Richard Florida (2005) once referred to as the creative class (*Regionalpolitiska utredningens slutbetänkande* 2000). The problem highlighted by critics is that, apart from in the metropolitan regions, it is difficult to create the kind of attractiveness that appeals to the capital that is required to compensate for the loss of previous regional development support from the government. As a result of this shift in Swedish regional development policy, also referred to as the "Superstar policy" (Forsberg and Lindgren 2010), transport planning institutions at the regional level currently support policies that encourage investment in regional commuter train services to merge the regions with larger and intraregional functional labour markets. In Sweden, the priorities of fast regional commuting have become a contributing factor to people relocating, despite the effects this might have on family life (Eliasson et al. 2007). Over the last two decades, long-distance commuting has become more common, especially among well-educated men. The policy of enlarging functional regions in Sweden has been criticized by feminist scholars (Friberg 2008; Lindkvist Scholten and Jönsson 2010; Gil Solá 2013) for viewing the commuter as a disembodied and free-floating individual with no family responsibilities or other attachments. From a feminist perspective, the top-down approaches to structural problems are in line with what

Kębłowski and Bassens (2018) describe as neoclassical planning practices. Erika Sandow scrutinizes the outcomes of long-distance commuting and its impacts on households and relationships. In her chapter, Sandow shows that personal relationships become stressed when a householder has to commute. In addition, she adds that women's long-distance commuting is not as easily accepted as men's from a normative standpoint regarding family and social network responsibilities. However, families that continue living in the same place for the first few years of commuting seem to organize family life in such a manner as to cope with the situation, which keeps the relationship going.

Women's and men's differing household responsibilities are also examined in the next chapter. Using a case from a Spanish travel survey, Inés Sánchez de Madariaga and Elena Zucchini identify that most trips made by women are what they define as care trips related to the upkeep of daily life: taking children to and from school or leisure activities, non-leisure shopping, errands to private and public offices, accompanying elderly relatives and so on. By examining more detailed levels of the statistics, including those on trips shorter than 15 minutes, which are invisible in aggregated national statistics, the authors reveal the number of women's trips that are related to what they refer to as mobilities of care. A reconceptualization and reframing of existing data in this way can provide knowledge about how women (still) have the primary responsibility for care trips, while also highlighting the important methodological implications that the gendering of conventional travel surveys has and must have.

WAYS FORWARD IN TRANSPORT PLANNING: MOVING BEYOND GENDER STEREOTYPES

Although analyses concerning men and women are still very pertinent and necessary, the need to differentiate not only along axes of gender but also along those of age, class, ethnicity and other power orders have been raised by numerous researchers. The fourth part of this collection addresses how transport planning needs to move beyond binary gender categories and analyse how the complexity of intersections manifests in people's everyday lives. Malin Henriksson focuses on sustainable transport planning in her chapter and discovers that everyday planning is a messy business in which the goals sometimes become contradictory or lack the appropriate tools regarding how to implement the policy objectives.

Henriksson learns that, because transport planners lack in-depth knowledge of how to investigate and analyse user groups' preferences, they revert to their own personal decision-making regarding modes of transport as a strategy for understanding travel behaviour—referred to as 'the I-methodology'. The problem Henriksson identifies is that, by using the I-methodology, transport planners' ideas about the social complexity of transport system users are overlooked; in addition, the users are constructed as middle-class, middle-income and, generally, well-educated and informed citizens, much like the transport planners themselves. The analysis that transport planners make to support the modal shift towards more sustainable modes of transport tends to be based on their own travel choices. To counter this, Henriksson calls for an intersectional analysis in which differences are acknowledged, and the gendering of space and the distribution of everyday destinations and amenities are analysed using concepts of gendered power structures in society.

In the interviews upon which Henriksson builds her analysis, the transport planners discuss the importance of teaching the young generation the significance of sustainable transportation. The men, in particular, describe how they take their children cycling to get to school or kindergarten, or just for fun in order to teach them how to become environmentally friendly transport users. This can be seen as counteracting the pervasive sociocultural discourse of the private car as the superior mode of transport. Globally, the iconic status of the car as the ultimate mode of transport since the mid-twentieth century has led to a situation in which walking, public transport and cycling have become deprioritized in many countries. Planning for the private car has become so internalized within the urban fabric that politics has not identified its dominant role in the designing of transport systems and transport environments (Koglin 2015). Space has been transferred from walking and cycling to car use, and bicycle lanes have been laid out alongside automobile thoroughfares rather than taking faster and shorter routes. In this collection, Dag Balkmar analyses how the everyday trips of cyclists in the contemporary transport system can be described, at their worst, as a battle during which they—as vulnerable transport subjects in a dangerous transport environment—are subjected to violent acts by motorists. Balkmar investigates the discourse of men as cyclists and, using data from an online discussion on cycling by cyclists, interviews with cyclists and promoters of cycling, and newspaper stories on cycling, he focuses on the modal conflicts in the transport system. The analysis of the data leads to three discourses regarding men, cycling and modal conflicts: men as the solution to sustainable transport, men as the problem and men as vulnerable transport subjects. Here again, an important conclusion of the chapter is that transport planners need to open up to a more complex understanding of the subject in order to produce a safer cycling transport environment and to work towards the political objectives of sustainability.

Tanja Joelsson focuses on children in her chapter. Building on research that denotes children as competent individuals with the ability to make decisions that favour themselves, Joelsson claims that children are capable of using the transport system if they are acknowledged by society as rightful users. The right to move and travel independently in and around a neighbourhood also has positive effects on children's development, health and wellbeing. However, Joelsson finds that children are poorly recognized in transport policy. By investigating local and regional transport planning policy, she deconstructs the idea of children in these policies and finds two distinct discourses, in which children are either absent or constructed as particularly worthy of protection. This leads to a discussion on children as citizens and their rights and access to public space and the transport system. Joelsson raises the question of how to reshape the planning process without having a romanticized idea of children's contribution to it. When everyday life is brought centre-stage, the spaces around which children's lives revolve become highlighted. Thus, it becomes possible to reformulate the political in transport policy and planning, producing space where the political is performed, as Joelsson puts it. This would not only enable children's and young people's appropriation of public space and create safer and healthier transport systems and environments, but it would also benefit everyone who lives in urban regions.

The last contribution in the section on moving beyond gender binaries in transport planning is written by Hilda Rømer Christensen, who develops an analysis of the gendered mobilities of Chinese cyclists. In this iconic cycling country—where the bicycle represented modernity as well as gender equality during the Mao era—economic development has led to a dramatic increase in the number of privately owned cars. By adopting a cultural assemblage methodology, Rømer Christensen presents examples of the power of global neoliberal economics. She dismantles the previous gender-equal policy and demonstrates how Western influences impact not only upon the multiplicity of gender identities but also upon modal preferences. This is achieved through her exploration of the political shift towards market liberation and individualization. Both national and global neoliberal politics, which are important in Rømer Christensen's argument, also affect local transport planning in relation to growth-oriented and neoliberal politics in regional development.

In the concluding chapter, we revisit the findings of the collection's contributions in the light of urgent concerns about mobility justice and environmental sustainability. We argue, in line with Sheller (2012), that the processes of social equality and sustainability are intimately entwined. This calls for bold and innovative interventions, both short and long term, and, strengthened by the collection's diverse and novel approaches, we conclude the chapter and the collection by making some suggestions for what we perceive as ways forward. Overall, we hope that this collection will inspire and nurture the debate surrounding feminist research and discussions on gender, power democracy, and the priorities that are being set within the transport research field. It is our firm conviction that the much-needed transformation of the transport system—in order to conform to global initiatives for a more socially just and environmentally sustainable future—must integrate dimensions of power, citizenship and democracy into policy-making within transport and transport planning practice. We believe there is still time to make change happen. Moreover, we believe in the power of politics. Finally, we believe in open and democratic processes as a way forward towards a more feminist-inspired understanding of the political field of transportation and mobility.

References

Bacchi, C. L. (2009). *Analysing policy: What's the problem represented to be?* Frenchs Forest: Pearson.

Balkmar, D. (2012). *On men and cars: An ethnographic study of gendered, risky and dangerous relations.* Dissertation, Linköpings Universitet, Linköping.

Banister, D. (2008). The sustainable mobility paradigm. *Transport Policy, 15*(2), 73–80.

Barker, J., Kraftl, P., Horton, J., & Tucker, F. (2009). The road less travelled: New directions in children's and young people's mobility. *Mobilities, 4*(1), 1–10.

Berg, J., Levin, L., Abrahamsson, M., & Hagberg, J. (2014). Mobility in the transition to retirement: The intertwining of transportation and everyday projects. *Journal of Transport Geography, 38*, 48–54.

Brown, B., Mackett, R., Gong, Y., Kitazawa, K., & Paskins, J. (2008). Gender differences in children's pathways to independent mobility. *Children's Geographies, 6*(4), 385–401.

Butler, J. (2006). Performative acts and gender constitution: An essay in phenomenology and feminist theory. In *The RoutledgeFalmer reader in gender & education* (pp. 73–83). London & New York: Routledge.

Cockburn, C., & Omrod, S. (1993). Gender in the microwave world. In *Gender and technology in the making* (pp. 41–74). London: Sage.

Dobbs, L. (2007). Stuck in the slow lane: Reconceptualizing the links between gender, transport and employment. *Gender, Work & Organization, 14*(2), 85–108.

Eliasson, K., Westerlund, O., & Åström, J. (2007). *Flyttning och pendling i Sverige.* Stockholm: Fritze.

Fainstein, S. S., & Servon, L. J. (2005). *Gender and planning: A reader.* New Brunswick: Rutgers University Press.

Florida, R. (2005). *Cities and the creative class.* New York: Routledge.

Forsberg, G., & Lindgren, G. (2010). *Nätverk och skuggstrukturer i regionalpolitiken.* Karlstad: Karlstad University Press.

Friberg, T. (2008). Det uppsplittrade rummet: regionförstoring i ett genusperspektiv. Regionalpolitikens geografi: regional tillväxt i teori och praktik. In F. Andersson, R. E., & I. Molina (Eds.), *Regionalpolitikens geografi: regional tillväxt i teori och praktik* (pp. 257–283). 1. uppl. Lund: Studentlitteratur.

Fyhri, A., Hjorthol, R., Mackett, R. L., Fotel, T. N., & Kyttä, M. (2011). Children's active travel and independent mobility in four countries: Development, social contributing trends and measures. *Transport Policy, 18*(5), 703–710.

Gil Solá, A. (2013). *På väg mot jämställda arbetsresor: vardagens mobilitet i förändring och förhandling.* Dissertation, Göteborgs Universitet, Göteborg.

Hanson, S. (2010). Gender and mobility: New approaches for informing sustainability. *Gender, Place & Culture, 17*(1), 5–23.

Hanson, S., & Pratt, G. J. (1995). *Gender, work, and space* (1st ed.). London [u.a.]: Routledge.

Healey, P. (2006). *Urban complexity and spatial strategies: Towards a relational planning for our times.* London: Routledge.

Hemmings, C. (2005). Telling feminist stories. *Feminist Theory, 6*(2), 115–139.

Hull, A. (2008). Policy integration: What will it take to achieve more sustainable transport solutions in cities? *Transport Policy, 15*(2), 94–103.

Joelsson, T. (2013). *Space and sensibility: Young men's risk-taking with motor vehicles.* Dissertation, Linköpings Universitet, Linköping.

Kębłowski, W., & Bassens, D. (2018). "All transport problems are essentially mathematical": The uneven resonance of academic transport and mobility knowledge in Brussels. *Urban Geography, 39*(3), 413–437.

Koglin, T. (2015). Velomobility and the politics of transport planning. *GeoJournal, 80*(4), 569–586.

Latour, B. (1993). *We have never been modern.* New York/London: Harvester Wheatsheaf.

Law, R. (1999). Beyond 'women and transport': Towards new geographies of gender and daily mobility. *Progress in Human Geography, 23*(4), 567–588.

Legacy, C. (2016). Transforming transport planning in the postpolitical era. *Urban Studies, 53*(14), 3108.

Levin, L. (2008). Äldre kvinnor: osynliga i statistiken men närvarande i trafiken. In M. Brusman, T. Friberg, & J. Summerton (Eds.), *Resande, planering, makt* (pp. 23–40). Lund: Arkiv.

Levin, L., & Faith-Ell, C. (2011). Women and men in public consultations of road-building projects. *Transportation Research Board Conference Proceedings.*

Levin, L., Faith-Ell, C., Scholten, C., Aretun, Å., Halling, J., & Thoresson, K. (2016). *Att integrera jämställdhet i länstransportplanering: Slutredovisning av forskningsprojektet Implementering av metod för jämställdhetskonsekvensbedömning (JKB) i svensk transportinfrastrukturplanering.* Lund: K2.

Levy, C. (2013a). *Transport, diversity and the socially just city: The significance of gender relations.* DPU, UCL & Universidad Nacional de Colombia (Medelln Campus).

Levy, C. (2013b). Travel choice reframed: 'Deep distribution' and gender in urban transport. *Environment and Urbanization, 25*(1), 47–63.

Lindkvist Scholten, C., & Jönsson, S. (2010). *Freedom of choice?: About long-distant commuters and employers approach to the effects of regional enlargement.* Växjö: Växjö University, Department of Social Sciences.

Listerborn, C. (2007). Who speaks? And who listens? The relationship between planners and women's participation in local planning in a multi-cultural urban environment. *GeoJournal, 70*(1), 61–74.

Lucas, K. (2004). *Running on empty: Transport, social exclusion and environmental justice.* Oxford/Chicago: Policy Press.

Lucas, K. (2012). Transport and social exclusion: Where are we now? *Transport Policy, 20,* 105–113.

Lykke, N. (2009). *Genusforskning – en guide till feministisk teori, metodologi och skrift.* Stockholm: Liber.

Lykke, N., & Braidotti, R. (Eds.). (1996). *Between monsters goddesses and cyborgs: Feminist confrontations with science, medicine and cyberspace.* London: Zed.

Martens, K. (2016). *Transport justice: Designing fair transportation systems.* London: Routledge.

Matrix (Organization). (1984). *Making space: Women and the man-made environment.* London: Pluto Press.

May, A. D., Page, M., & Hull, A. (2008). Developing a set of decision-support tools for sustainable urban transport in the UK. *Transport Policy, 15*(6), 328–340.

Mouffe, C. (2005). *The return of the political.* London: Verso.

Mouffe, C. (2013). *Agonistics: Thinking the world politically.* London: Verso.

Oliver, K. (1988). Women's accessibility and transport policy in Britain. In S. Whatmore & J. Little (Eds.), *Gender and geography* (pp. 19–34). London: Association for Curriculum Development.

Regionalpolitiska utredningens slutbetänkande. (2000). Stockholm: Fritzes offentliga publikationer.

Rosenbloom, S. (2005). Women's travel issues. In S. S. Fainstein & L. J. Servon (Eds.), *Gender and planning: A reader.* New Brunswick: Rutgers University Press.

Rubin, G. (1975). The Traffic in women. In R. R. Reiter (Ed.), *Toward an anthropology of women* (p. 157). New York: Monthly Review Press.

Salon, D., & Gulyani, S. (2010). Mobility, poverty, and gender: Travel 'choices' of slum residents in Nairobi, Kenya. *Transport Reviews, 30*(5), 641–657.

Scholten, C., Friberg, T., & Sandén, A. (2012). Re-reading time-geography from a gender perspective: Examples from gendered mobility. *Tijdschrift voor Economische en Sociale Geografie, 103*(5), 584–600.

Schwanen, T., Kwan, M., & Ren, F. (2008). How fixed is fixed? Gendered rigidity of space–time constraints and geographies of everyday activities. *Geoforum, 39,* 2109–2121.

Sheller, M. (2012). Mobilities. In *The Wiley-Blackwell encyclopedia of globalization.* Chichester/Malden: Wiley Blackwell.

Uteng, T. P., & Cresswell, T. (2008). Gendered mobilities: Towards an holistic understanding. In T. P. Uteng & T. Cresswell (Eds.), *Gendered mobilities* (pp. 15–26). London/New York: Routledge.

Wajcman, J. (2010). Feminist theories of technology. *Cambridge Journal of Economics, 34*(1), 143–152.

West, C., & Zimmerman, D. H. (1987). Doing gender. *Gender & Society, 1*(2), 125–151.

Wolff, J. (1993). On the road again: Metaphors of travel in cultural criticism. *Cultural Studies, 7*(2), 224–239.

PART I

Feminist Interventions
in Transport Planning

CHAPTER 2

Are We Still Not There Yet? Moving Further Along the Gender Highway

Clara Greed

Introduction

In this chapter, I want to reflect on key themes and conclusions expressed by the authors of the various chapters, about how to integrate gender awareness into transportation policy. I will do this with reference to my own research on similar issues in the UK. As discussed in an earlier chapter (Greed 2012) (upon which this chapter draws), the sustainability agenda has been seen by those seeking "a way in" to the still male-dominated world of transport planning as a pathway to integrating the so-called social considerations (are not all policy issues social?) into transport policies. In particular, feminist transport experts have investigated the potential and limitations of this approach, as discussed in this book. Sustainability, especially environmental sustainability, has been a key driving force in planning policy for nearly 30 years (Bruntland Report 1987). However, many people, especially women, have found that environmentally focused sustainability policies have made their daily lives more difficult, whilst not

C. Greed (✉)
Faculty of Environment and Technology,
University of the West of England, Bristol, UK
e-mail: clara.greed@uwe.ac.uk

© The Author(s) 2019
C. L. Scholten, T. Joelsson (eds.), *Integrating Gender into Transport Planning*, https://doi.org/10.1007/978-3-030-05042-9_2

25

necessarily enabling them to adopt a greener lifestyle. There appears to be a lack of consideration of the differential social impact of physical transport policy on different social groups, especially women, and other relatively powerless and unrepresented groups. There has been an over-emphasis upon both the environmental and technical aspects of transport planning, at the expense of social considerations, especially gender. Examples from the various chapters of this book show that gender-ignorant transport planning policy is working against the creation of inclusive, equitable and accessible transport systems. In this chapter, I will again examine 'the problem' in two stages: firstly, regarding the inherited 'pre-sustainability' nature of British cities and, secondly, the challenges the 'sustainable city' creates for its citizens. For example, in the past, transport planners seemed to have tremendous power to demolish large sectors of cities, because, as technocrats, they positioned themselves as being 'above' political accountability and normal democratic processes. Nowadays, sustainability policy often seems to be equally unaccountable, because of its assumed inviolable holiness, such as the unquestioned worthiness of the environmental movement. But policy is often set at too high a level to recognize or engage with the realities of everyday life.

CONCEPTUAL AND METHODOLOGICAL PERSPECTIVES

Previously, I have set out the issues and possible remedies in a chapter entitled, "Are we there yet: women and transport revisited", which I wrote for another landmark book, *Gendered Mobilities* (Greed 2008; Uteng and Cresswell 2008). In subsequent publications, I have sought to raise awareness of gendered transport issues among professionals, academics (Greed 2012) and students (Greed and Johnson 2014, Chapters 12 and 15). But, some of the contributors to this book have taken the whole debate much further, for example, into the realms of urban politics, new methodologies, new research topics and innovative solutions.

Much of my research has been concerned with investigating different aspects of the urban question, namely, "who gets what where why and how?" (Harvey 1975) with particular reference to women, and likewise many of the chapters in this book provide answers and examples in relation to transport planning. I have found that a key determinant in shaping cities is the nature of planning policy, and thus the perspective and 'worldview' of the policy-makers themselves (Greed 1994a: 10). I have been concerned with the 'dissonance' (contrast and incompatibility) between what the planners imagine is required and the realities experienced by the urban population as they seek to access and use 'the city of everyday life'. Much policy seems to be set at too high a level, or to relate to generalizations and stereotypes, which do not relate to the lives of ordinary people. As pointed out by various contributors to this book, this situation arises because of deeper political issues around the concepts of democracy and the representation of neglected groups within society and why their voices are not heard (Mouffe 2005).

Conceptually, my research has been concerned with the reproduction across space of social relations, including the imprint of gender relations on the built environment (Massey 1984: 16). The problem of the male-dominated nature of urban planning, and of transport planning, in particular, has been widely researched and documented over many years (Stimpson et al. 1981; Matrix 1984; Little et al. 1988; Whatmore and Little 1988; Roberts 1991; Greed 1994a; Booth et al. 1996; Buckingham-Hatfield 2000; Anthony 2001; Hayden 2002; Reeves 2005; Uteng and Cresswell 2008; Jarvis et al. 2009; inter alia).

In particular, I have found there to be little recognition of the social dimension of sustainability (Vallance et al. 2011), which so strongly shapes transport policy nowadays. In particular, the needs of women seem to be completely left out of much so-called sustainable transport policy (Lucas 2012). But there is little data on women's "different" travel needs and patterns (CIC 2009). Therefore, the spatial and transport needs of women, and other minority groups, are unlikely to be taken into account within the professional psyche of the predominantly male transport planning professions, or seen as being of any importance.

Much of my own research has taken a sociological, even anthropological, approach to investigating the sub-cultural values of the planning profession, and the ways in which culture has rendered it ignorant and often hostile towards the 'different' needs of women (Greed 1994a, 2011, 2012; Greed and Johnson 2014). I have generally used qualitative, ethnographic approaches (Greed 1994b) in order to make sense of the policy stances and lack of reference to social inclusion considerations. But I have also been involved in more quantitative research too. For example, with Dory Reeves, I undertook a national study in the UK for the Royal Town Planning Institute (RTPI) on the extent to which gender considerations were being mainstreamed into spatial policy in local planning authorities (RTPI 2003; Greed 2005).

However, in this book, several of the contributors have undertaken much more quantitative and statistical research using mathematical and scientific methodologies which are more commonly found in, and more acceptable to, the male-dominated world of transport planning (including Lena Smidfelt Rosqvist, Erika Sandow, Inés Sanchez de Madariaga and Elena Zucchini) to beat the men at their own game. In addition, Malin Henriksson in her chapter describes how she applies the innovative I-methodology when analysing how transport planners create and evaluate the 'sustainable traveller', basically founded upon the transport planners' own preferences. Some have investigated the minutiae of what at first sight may appear to be rather unpromising topics for gender research, such as driving tests and the choice and costing of construction materials (Lena Smidfelt Rosqvist). In this case, the contributor has actually filled in new pieces of the jigsaw, and helped us to realize the huge impact that topics within which one would imagine gender to be irrelevant actually have within the transportation industry on minimizing women's voices. In contrast, other contributors (Karen Levy, Tanja Joelsson, Lena Levin and Charlotta Faith-Ell) have taken a wider, more political, philosophical and sociological perspective on the issues (e.g. in Part I of this book). They have identified the deeper constraints that result in the lack of political power and policy-making influence amongst so-called minority groups in our society, including women, children and the elderly: who together actually make up the majority of society. Thus, they have taken the debate into the realms of power in society, and have drawn upon issues of democracy and representation.

DEFINITIONS: SUSTAINABILITY AND INCLUSIVE DESIGN

The need to create sustainable cities is a key driving force in spatial policy-making. But I have found, in the UK, that a partial and incomplete approach to sustainability is used that over-emphasizes the environmental dimension at the expense of social considerations. The original definition of sustainability in the Rio Declaration (Buckingham-Hatfield 2000) included three components: economic viability, social equity and environmental sustainability (prosperity, people, place). In other European countries, especially within Scandinavian countries, the social component is given greater importance (Madariaga and Roberts 2013). But, in Britain, a somewhat 'peopleless' concern with 'green' environmental issues appears to have eclipsed the social component, detracting from diversity considerations, and thus reducing the chances of achieving fully sustainable cities. Friends of the Earth has declared that "transport is one of the worst perpetrators of sexual discrimination", but the social dimension seems to have been jostled out of the way in the modern-day environmental UK sustainability agenda (Oliver 1988). 'Inclusive urban design' and 'sustainability' have become disconnected, especially with respect to women's issues. Inclusive urban design may be defined as an approach to the planning, design and layout of our towns and cities that recognizes and accommodates the needs of those with disabilities and benefits all those other societal groups, especially women, that are currently dis-enabled by the nature of the built environment. Inclusive urban design would create enabling environments in which, for example, street layouts would be accessible, functional and direct, as well as being safe, attractive, legible and easy to negotiate (Manley 1998; Imrie and Hall 2001). Such an approach would extend the principles of universal and inclusive design (Goldsmith 2000) to the planning of entire cities in terms of strategic policy-making affecting the nature of transport systems, zoning policy and policy priorities in terms of the location, distribution and design of the different spatial components, land uses, transport systems and other spatial components and amenities that make up the urban environment (Anthony 2001).

The 'Old' Problem: The Unsustainable City of Man

Historical planning policy has resulted in cities being based upon the zoning and separation of land use, creating unnecessary distances between home and work and thus unsustainable commuting patterns. Post-war UK planning prioritized housing clearance and the dispersal of industry, residential areas and public facilities, thus endorsing low-density, suburban development around our cities and creating extended work commuting distances and enclosed housing estates for women (Roberts 1991: 70–77). Subsequent governments favoured car use, and the 'Americanization of British cities', although North American women planners had already warned against the problems of urban decentralization, over-zealous zoning and car-dominant cities (Stimpson et al. 1981). Erika Sandow, in her chapter, discusses the problems that long-distance commuting creates, which has arisen, in part, because of urban decentralization and the disconnection of work and home localities. From the 1960s onwards, entire districts of cities were bulldozed to make way for urban motorways and car parks for the predominantly car-borne male commuter. Thus, a car-based urban infrastructure, with dispersed land-use patterns, was developed, whilst public transport was left to decline. By the end of the twentieth century, a range of developments, including out-of-town shopping malls, hypermarkets, business parks and leisure facilities, have become located alongside the motorway system, whilst schools, hospitals and local authority offices have all been decentralized in the name of efficiency, thus undermining the viability of existing city centres and traditional towns in the process. All these policies have increased traffic congestion and journey times as people try to travel between different land-use areas to carry out their daily tasks. This is so much the case that it is now a virtual necessity to possess a motor car, particularly in cities where public transport has been severely cut back and land uses are highly dispersed, with essential facilities, shops and amenities no longer within walking distance. Women have been particularly adversely affected by these changes because women and men still have different roles and responsibilities, and therefore different travel patterns within the city of man (Anthony 2001).

Women have been poorly served by planning policies that do little to recognize or plan for their "different" needs and travel patterns in the development of land-use policy (Coleman 2008; Reeves 2005), as illustrated in Part II on 'Everyday Mobility'. For example, in her chapter, Lena Smidfelt Rosqvist admirably puts forward the arguments and constraints limiting women's travel choices. Women are still predominantly the ones responsible for childcare, shopping and home-making, although the majority of women also work outside the home, and for all the other 'care journeys' which Elena Zucchini and Inés Sanchez de Madariaga discuss in their chapter. This means that women's 'journey to work' is very complex and multi-purpose compared with the simple, uninterrupted, mono-journey to work and back again of the traditional male commuter, and far more difficult to achieve in cities that are still structured around past zoning priorities and decentralization. Therefore, trip-chaining and multi-tasking are key features of women's travel, and an inevitable result of trying to combine their home and work. For example, a woman may set off from home, stop at the childminder's, then school, get to work, and return via the school gates, shops and childminder, resulting in complex trip-chaining. This daily travel itinerary is difficult to achieve when employment and residential areas have been separated out by traditional land-use zoning, based upon male perceptions of spatial functionality (Uteng and Cresswell 2008), whereas for women a range of more personal factors come into play, such as personal safety, crime, street lighting, pavement condition, accessibility and practicality (Booth et al. 1996). Women's journey needs are already poorly met by public transport systems that have been designed on a radial basis, funnelling workers into the centre during rush hour and providing limited off-peak services for women and part-time workers and a lack of transverse, inter-district bus routes. More women than men work part-time or hours that are outside the traditional nine to five of office workers. For example, before the rush hour starts, early-morning cleaners have already worked several hours in offices, and factory workers have already left early and commuted out to factories at the city periphery. In order to enable women and men of all ages to travel comfortably and easily, it is important to make transport systems accessible and usable, with adequate ancillary facilities.

THE 'NEW' PROBLEM: THE SUSTAINABLE CITY

Far from solving the 'old' problems described above, the new sustainability-orientated planning agenda is compounding these problems and creating 'new' problems of its own, not least because of a lack of a gender perspective on planning, and an overall lack of awareness of social user needs or the realities of everyday life. Solutions appear to be focused upon restriction, control and penalization, or upon a condemnation of personal lifestyle choices, without offering alternatives based upon investment in structural spatial change and better transport systems and services. In spite of social class being part of the diversity agenda, one senses a certain contempt for 'the people' within the sustainability agenda. But individuals still have to get to work; they are not free agents, and will have no choice but to use their cars if no other alternative is available. In particular, many women are time-poor and the only way to carry out the elaborate trip-chaining journeys described above is by car. After many years of promoting the motor-car, planners have made a *volte-face* and now condemn car use rather than promoting it, in the name of sustainability, but they still retain the same power to shape cities. Policy proposals to restrict the motorcar and encourage public transport use might at first have been welcomed by the public as a way of easing congestion and making cities more efficient. However, these restrictions were not accompanied by corresponding measures to increase public transport provision. There appears to be little appreciation within the government of the necessary investment and infrastructural development, and of all the ancillary preparations and facilities that must be put in place to enable the majority of the population to travel 'en masse' by public transport. A new generation of young, green transport planners has arisen who are fired up with environmental fundamentalism. They justify their actions in the name of sustainability and are apparently above reproach because of their zeal for the environment and respect for the Planet. If one dares to ask, "what about women?" one is likely to be told (as I have been many times), "oh we've done women, you should be concerned about the environment". There seems to be little understanding of the complex and essential nature of people's journeys, especially women's, and the fact that public transport does not provide the routes, destinations or timetable provision that women and men require. As a result, women travel around cities—with difficulty—using their survival skills and precise knowledge of local bus timetables, with limited time budgets to get everything done. Public transport is inadequate, expensive, unreliable and infrequent.

THE NEW EMPHASIS ON PUBLIC TRANSPORT

It is impractical to suggest that everyone should get back onto public transport because the services have already deteriorated (Hamilton et al. 2005) and are barely meeting people's existing needs, when only a minority are using them. Some groups have never had cars in the first place and find their needs displaced in favour of attracting car-driving commuters onto public transport. Existing local bus routes that went around residential areas are being cancelled or diverted to provide buses to serve the more direct routes from Park and Ride car parks to the city centre (as in Brislington, Bristol). Likewise, the urban Tramway in Croydon, South London, has improved speed of travel within the town centre but has resulted in existing bus routes coming in from the suburbs being diverted and travel times increased (Greed and Johnson 2014: 240). Much of the hype about people adopting more sustainable lifestyles gives the impression that public transport is readily available and accessible to all. However, there are large areas of the country, particularly outside London, where public transport barely exists, with, at most, infrequent and unreliable bus services. The majority of the population lives in suburban areas, comprising both decentralized council housing estates and private residential developments, and many of these areas, particularly in the provinces, lack adequate and reliable public transport systems, and have never had access to the rail system. Many small towns and villages had their railway stations closed down as a result of the Beeching railway cuts (Beeching 1963) and promises of alternative public transport never materialized (Greed and

Johnson 2014, Chapter 12). To save time, trains may not stop at intermediate stations because of the emphasis on inter-city commuter routes. Therefore, of necessity, many people resort to using cars for the journey to work, because there is no alternative. Penalizing the motorist does not give individual citizens the power to re-open railway stations, to create new bus routes or to change ingrained land-use patterns.

THE SCHOOL RUN AND RELATED JOURNEYS

Children's transport needs often seem to be either ignored or treated as a problem and inconvenience, and they (and their accompanying parents) are often not seen as being 'entitled' to travel or to have their needs valued in policy-making, as discussed in the chapter by Tanja Joelsson. Particular hatred seems to be reserved for 'the school run', an activity predominantly undertaken by women taking their children to school by car and parking outside the school. This activity only contributes to about 15% of rush-hour traffic, but it is widely condemned for "cluttering up the roads" (Hamilton et al. 2005). In contrast, no-one criticizes the 'office run' or the congestion caused by husbands being dropped off at the railway station by their wives. The school run is portrayed in the media as being undertaken by rich lazy housewives in their 4-by-4 Range Rovers, although many families only own a cheap car and have to make major economies to keep it running. (Seldom does one hear that some male parents do in fact undertake the school run too.) One feels a sense of *deja vu*, recalling previous generations of transport planners who condemned 'women car drivers' and their essential journeys (for work, school, shopping and childcare) as leisure journeys that got in the way of the journeys of the male breadwinner. Greener alternatives are promoted, such as the 'walking train', where school children parade along in a crocodile to school with mothers being encouraged to supervise this perambulation. Such schemes assume that mothers will be available early in the morning. In reality, many women are 'time poor' and very anxious at this time of day as they are frantic to get to work. One of the most efficient ways of ensuring that their children get to school on time is to drop them off on the way to work, as part of the morning trip-chain, with mothers often sharing this role. 'Walking-trains to school' schemes might be viable in higher-density urban areas but are hardly practical in spread-out suburban locations. However, teachers or official wardens, not mothers, should be provided to staff the activity and should be remunerated for doing so. The journey to school is underestimated, in terms of both the numbers involved, and the time and commitment provided by parents to ensure their children get to school on time. Local authorities seem surprised when they find large numbers of children, and for that matter cars, converging on the school gates in the morning and their reaction is largely negative, fining people for parking and condemning the activity. But women are by default providing 'public transport' in their private cars, compensating for the lack of government provision. Many of the supportive journey activities undertaken by parents are unrecognized, unpaid and condemned, such as undertaking 'escort' journeys to ferry children and teenagers around in the evenings because public transport is so limited, and parents are wary of letting their children out on their own.

Many main roads into cities provide a priority lane for those with two or more passengers, with cameras checking the number of passengers. Although mothers may have ferried several other people and children in their car in the course of their morning trip-chain, by the time they head for the final stretch to work they may be on their own again and find they are not entitled to use the 2+ lane and, although arriving later than others, are also not entitled to use the 2+ parking spaces. Both of these contriv-ances favour the rush-hour commuter on his uncomplicated and unbur-dened 'journey to work'. Parents, who by their unselfish and complicated car journeys are constantly compensating for the inadequacies of urban form, planning policy and public transport provision, are likely to be penalized by these measures. Road charging, for example the 'Congestion Zone' in central London, is a crude way of promoting sustainability, based upon the ability to pay, not upon the usefulness of the journeys under-taken. It can cut into women's trip-chaining if they have to pass in and out of the zone several times during the day as they undertake their sequence of trips to and from the childminder, for example. But public transport also presents insurmountable difficulties. If mothers, while taking children to school, try to use the bus, train or Tube in the rush hour, they are likely to get 'condemnatory looks' for 'cluttering up the public transport sys-tem' with their offspring. It is illegal to leave small children unattended at home, so pre-school children and most likely their pushchairs will have to come along for the journey too. Some escalators still have signs banning pushchairs but, at the same time, women may be shouted at for using the disabled lift (Lenclos 2002). Public transport is far from 'public', and, as the Consumer Association commented years ago, is mainly aimed at able-bodied men "carrying nothing more than a rolled-up newspaper".

PROVISION OF ANCILLARY FACILITIES AND LOCAL CENTRES

The promotion of cycling and walking is of value for short local journeys, but expecting everyone to stop using their cars ignores the long distances that many people have to commute to get to work, which in turn are the result of past land-use zoning and decentralization policies. The closure of local shops and decentralization of retail stores also makes it increasingly difficult to food-shop without using a car. Indeed, as discussed by Dag Balkmar in his chapter, cycling may be seen as a solution to many car-based transport problems, but it can also be part of a new set of problems, particularly for pedestrians, who often feel endangered by thoughtless cyclists. In particular, women with small children often feel threatened by certain male cyclists who express their machismo by imagining they are competing in the Tour de France as they speed carelessly through busy pedestrian areas. These problems are becoming particularly marked where policies of 'shared space' are being pursued without adequate thought, which expect bicycles, pedestrians and even cars to share the same road space, with the intention of reducing car speeds and making streets 'safer'. But such approaches do not take into account gender considerations, such as the fact that women often feel intimidated about 'claiming their space'. Thus, a man might stride out confidently in front of a moving vehicle knowing he will be respected and that it will stop for him, whereas a woman is more likely to be shouted at by the driver. But the gender dimension is completely missing from much shared-street literature (Hamilton-Baillie 2008).

There is clearly a gender dimension to cycling which is not adequately recognized; or, if it is, it is more likely to be related to male drivers harassing female pedestrians, rather than the problems that some male cyclists can create. But gender factors run deep, as discussed in Karen Levy's chapter on 'deep distribution' and gendered travel choices. Twice as many men as women cycle in Britain (WDS 2005). Deterrents include abuse from male drivers, personal safety fears on cycle paths, road safety concerns and lack of ancillary amenities. Buses, trains, railway and bus stations, pavements, streets, toilets and public spaces need to be accessible to everyone if the government is serious about creating sustainable cities. It is hypocritical to condemn people for using their cars if they are unable to access the public transport system because of steps, steep gradients, narrow footpaths obstructed by posts, poles and bins, poor lighting and unsafe layouts. These issues affect millions – not 'just' the disabled – but everyone who is dis-enabled by the design of the built environment, which includes anyone who feels vulnerable or finds the city difficult to negotiate. Much sustainability policy appears to be framed with little reference to the differences in travel patterns undertaken by women and men. Likewise, emphasis in existing inclusive design policy is generally upon facilitating accessibility at the 'micro' level of street layout, by remedial measures such as installing ramps, rather than promoting systemic change in land-use and transportation patterns at the city-wide strategic planning level, which would increase access and mobility for all. The British government has made much of creating 'sustainable communities' (ODPM 2003), and one would imagine this would be the 'magic link' between 'sustainability' and 'social inclusion'. However, there appears to be limited acknowledgement or understanding of the range of groups that comprise 'the community'. After the obligatory, introductory reference to the importance of gender, sex, race, age inter alia, 'disability' proves to be the main subject of much inclusive design guidance, thus obscuring other diversity issues that require attention (CABE 2008; TCPA 2009). Indeed, women are seldom mentioned in key policy documents on accessible and sustainable transport, resulting in the need for 'special' policy documents on 'women and transport' being produced, for example, by the Equal Opportunities Commission (Hamilton et al. 2005), and progressive local authorities (GLA 2007). A community is not sustainable if it ignores the needs of its elderly constituents (Age UK 2009: 44) and has limited understanding of women's 'different' use of the city and its transport systems (Uteng and Cresswell 2008) and only holds old-fashioned stereotypical views of men's activity patterns (Reeves 2005: 74). People over 50 years of age now constitute 21 million people (a third of the population) (Gilroy 2008), of whom over 65% are female, so clearly this is another 'women and planning' issue. Age UK emphasizes the need for a city-wide, strategic spatial approach to reshaping the city to meet the needs of the ageing population, stressing the key components of safe streets, transport, pavements, toilets, shops, places to meet, seating and so on. The link is made between sustainability and inclusive design by showing that an increased localization of food production, shops, housing and social facilities reduces climate change and the chances of local flooding, as well as diminishing the need to travel. Recommendations are not limited to ramps and disabled toilets, but address city-wide strategic planning policies on transport land use, and the location of local centres and facilities.

CONCLUSION AND RECOMMENDATIONS

The current approach to sustainability places disproportionate emphasis upon environmental factors, with particular concern about transportation issues. This makes it more difficult for people to live their lives, without providing viable alternatives, not least because women's journeys and land-use needs are not given adequate attention in sustainability policy. Rather than introducing negative car-controlling policies and tinkering around with traffic-light sequences and parking-space allocations, apparently to slow the traffic down and to discourage people from using their cars, there need to be positive measures, such as realistically investing in public transport to give people viable journey alternatives. In the longer term, fundamental structural changes are required in urban form and structure through forward planning policy. If the full agenda of 'sustainability' were taken into account, the social impracticalities would soon come to light. Likewise, if the full agenda of 'inclusive design' were taken into account, rather than restricting it to a few special disability measures, the needs of the majority of the population would be met, especially those of women and the elderly (ONS 2012).

Whilst much of the blame for a lack of inclusive urban design rests with town planners and urban designers, 'planning' is not all-powerful. The statutory powers governing the scope and nature of the planning system militates against the implementation of true sustainability policy because so many practical, everyday issues (such as toilets) and 'social' issues (such as childcare) are not 'officially' of concern to the planning system. Such matters are deemed *ultra vires* 'not a land-use matter' within the UK planning system. In contrast, environmental sustainability easily fits into the planning system, because it relates to the physical environment and the traditional protectionist concerns of town and country planning. The government did not hesitate to integrate EU directive requirements for Environmental Impact Assessment (EIA) into the planning system, but there is no parallel requirement that Social Impact Assessment (SIA) should be carried out. So, social considerations, unlike environmental requirements, do not generally feature in the main body of development plan documents. The contribution that a gender equality impact assessment can make to transport policy has been highlighted in the chapter by Lena Levin and Charlotta Faith-Ell with reference to Sweden, where it seems the situation is quite advanced compared with the UK and other European countries.

Nevertheless, in the UK, some progressive local authorities have included equality policies in the Supplementary Planning Guidance documents (GLA 2007) and have produced guidance reports on spatially relevant topics, such as public toilets, which fall outside the scope of statutory planning (GLA 2006; Greed 2003). I have argued in other publications (Greed 2016) that, if the government wants to create sustainable, equitable and accessible cities, then public toilets are the missing link. This is because people are not going to leave their cars at home and use public transport, especially for long journeys with small children, if there are no locally available toilets at transport termini and busy destinations. Women have more reasons to use the toilet than men, because of menstruation, yet this and many other bodily functions are never recognized as important in transport and urban planning policy (Greed 2016).

To achieve truly sustainable transport policy, there is a need to relate land-use patterns more directly to women's trip-chaining travel patterns and time-budgeting, and examples of this can already be found in some more progressive municipalities in Western Europe and North America (Fincher and Iveson 2008; Madariaga and Roberts 2013). For example, Groningen City Plan in the northern Netherlands requires that childcare provision is planned alongside school buildings to ensure that the trip-chain is simplified for busy parents, dropping off children on the way from home to work, thus making bicycle travel both possible and preferable. This policy is enabled by applying Dutch central government guidance on approaches to urban planning that combine work and family care needs within housing areas. Interestingly, there does not seem to be the same male macho culture associated with cycling in the Netherlands, perhaps because everyone cycles. Hilda Rømer Christensen has studied China in her chapter, a country that used to have a very high level of cycling activity based on gender-equal ideals, but which nowadays is facing the pressures and pollution of a massive growth in car ownership and a mobility paradigm where traditional gender ideals have been smeared into both car ownership and bike riding. Rømer Christensen shows how the Westernized, idealized gendered roles become challenged, and it will be interesting to see whether China can make cycling fashionable again without falling into the Western trap of it becoming a macho male lifestyle choice that impinges on everyone else's safety. Transport choice and policy, as Rømer Christensen explains it, can create new inequalities due to a lack of consideration not only of gender but of all the other social characteristics that intersect in people's lives, including age, social class and disability (Grabham et al. 2008).

Likewise, at the micro level of detailed street layout, for many years in Germany, there has been a concern to link gender issues with traffic management matters, such as traffic calming, (Hass-Klau 1990). Overall, there is little evidence of gender being taken seriously into account in British town planning (Greed 2005; Jarvis et al. 2009). Yet, the Gender Equality Duty (DCLG 2007) requires all local authority departments, including architecture and planning, not to discriminate in terms of policy-making, allocation of resources, provision of public services or recruitment and promotion of planning staff. The RTPI has published a *Toolkit for Practitioners* (RTPI 2003 and subsequent guidance), whilst the government itself has promoted diversity in planning (ODPM 2005).

Whilst we await such changes, there is a need to re-evaluate existing priorities and investment. If social and environmental issues were cross-referenced, then sustainable transport policy would prioritize different types of journeys as 'essential'. Rather than building new Park and Rides to meet the existing needs of predominantly male commuters, greater emphasis would be given to local and off-peak bus services to meet working women's unrecognized travel needs, and, as a result, the routes and timetables would be reconfigured to meet the substantial, unmet, 'off-peak' needs of the travelling public. As for car use, new car parks would be built, and not condemned outright, but aimed at meeting different priorities, with more flexible combinations of private and public transport modes. If sustainability policy were more 'joined up', collector car parks would be built around all suburban railway stations, especially in Greater London, and around bus termini in the main provincial towns. This would enable people to get to the railway station by car, and then park and use the train. En route to the new car parks, they could still carry out the other parts of their trip-chain, such as getting their children to school or childcare, and on the way back carry out essential food shopping and other necessary home-making duties. At present, many people use the car for their entire journey because there is no connecting bus route from their residential area to the nearest railway station (which may be 10 miles from home), and once in the car, it is quicker to go all the way. A wider picture of travel patterns and their social value would be built up. Supporting services such as toilets, bus-shelters, crèches, cycle lanes, steps, the storage and carriage of luggage, and shopping home delivery services would all be integral components of the transportation infrastructure. Nevertheless, some enlightened local authorities have sought to demonstrate the linkages between sustainability and social inclusion. For example, Plymouth first used a matrix model to link sustainability and gender issues in planning policy-making (Plymouth 2001). The Greater London

Authority and several of the London boroughs have strongly promoted the mainstreaming of gender into planning policy as well, though with variable results (GLA 2007). But much more needs to be achieved nationally.

To conclude, as an alternative to spread-out, zoned, low-density cities, many female planners would like to see the 'city of everyday life', which they define as the city of short distances, mixed land-use and multiple centres as the ideal objective that would fully take into account gender considerations as originally promoted by Eurofem (the European Women's Planning Network), and visionaries such as Liisa Horelli in Aalto, Finland (Eurofem 1998; Madariaga and Roberts 2013). Such a city structure would benefit all social groups, reduce the need to travel and create more sustainable cities that would be more accessible, whilst creating the higher-quality urban environments for all. This would provide more jobs and facilities locally and help to revitalize declining areas overall.

REFERENCES

Age UK. (2009). *One voice: Shaping our ageing society*. London: Age Concern.

Anthony, K. H. (2001). *Designing for diversity: Gender, race and ethnicity in the architectural profession*. Chicago: University of Illinois.

Beeching, R. (1963). *The Beeching report: Reshaping railways*. London: Her Majesty's Stationery Office.

Booth, C., Darke, J., & Yeandle, S. (1996). *Changing places: Women's lives in the city*. London: Paul Chapman.

Bruntland Report. (1987). *Our common future, World Commission on Environment and Development*. Oxford: Oxford University Press.

Buckingham-Hatfield, S. (2000). *Gender and environment*. London: Routledge.

CABE. (2008). *Inclusion by design: Equality, diversity and the built environment*. London: Commission for Architecture and the Built Environment.

CIC. (2009). *Gathering and reviewing data on diversity within the construction professions*. Construction Industry Council, London, in association with University of the West of England, Bristol.

Coleman, C. (2008). Women, transport and cities: An overview an agenda for research. In J. Darke, S. Ledwith, R. Woods, & J. Campling (Eds.), *Women and the city: Visibility and voice in urban space* (pp. 83–97). London: Palgrave.

DCLG. (2007). *Gender equality scheme*. London: Department of Communities and Local Government.

EUROFEM. (1998). *Gender and human settlements: Conference report on local and regional sustainable human development from a gender perspective*. Hämeenlinna: Eurofem.

Fincher, R., & Iveson, K. (2008). *Planning and diversity in the city: Redistribution, recognition and Encounter.* London: Palgrave Macmillan.

Gilroy, R. (2008). Places that support human flourishing: Lessons from later life. *Planning Theory and Practice, 9*(2), 145–163.

GLA. (2006). *An urgent need: The state of London's toilets.* London: London Assembly and Greater London Authority.

GLA. (2007). *Planning for equality and diversity in London, London: Greater London authority: Supplementary guidance to the greater London strategic development plan.* London: GLA (Greater London Authority).

Goldsmith, S. (2000). *Universal design: A manual of practical guidance for architects.* Oxford: Architectural Press.

Grabham, E., Cooper, D., Krishnadas, J., & Herman, D. (Eds.). (2008). *Intersectionality and beyond: Law, power and the politics of location.* London: Taylor and Francis.

Greed, C. (1994a). *Women and planning: Creating gendered realities.* London: Routledge.

Greed, C. (1994b). The place of ethnography in planning. *Planning Practice and Research, 9*(2), 119–127.

Greed, C. (2003). *Inclusive urban design: Public toilets.* Oxford: Architectural Press.

Greed, C. (2005). Overcoming the factors inhibiting the mainstreaming of gender into spatial planning policy in the United Kingdom. *Urban Studies, 42*(4), 1–31.

Greed, C. (2008). Are we there yet? Women and transport revisited. In T. P. Uteng & T. Cresswell (Eds.), *Gendered mobilities* (pp. 243–256). London: Ashgate.

Greed, C. (2011). Planning for sustainable urban areas or everyday life and inclusion. *Urban Design and Planning, 164*(2), 107–119.

Greed, C. (2012). Planning for sustainable transport or for people's needs. *Urban Design and Planning, 165*(4), 219–229.

Greed, C. (2016). Taking women's bodily functions into account in urban planning policy: Public toilets and menstruation. *Town Planning Review, 87*(5), 505–523.

Greed, C., & Johnson, D. (2014). *Planning in the UK: An introduction.* London: Palgrave Macmillan.

Hamilton, K., Jenkins, L., Hodgson, F., & Turner, J. (2005). *Promoting gender equality in transport.* Equal Opportunities Commission, Manchester, Working Paper No. 34.

Hamilton-Baillie, B. (2008). Towards shared space. *Urban Design International, 13*(2), 130–138.

Harvey, D. (1975). *Social justice and the city.* London: Arnold.

Hass-Klau, C. (1990). *The pedestrian and city traffic.* London: Belhaven Press.

Hayden, D. (2002). *Redesigning the American Dream: The future of housing, work and family life.* New York: Norton.

Imrie, R., & Hall, P. (2001). *Inclusive design: Designing and developing accessible environments.* London: Spon.

Jarvis, H., Cloke, J., & Kantor, P. (2009). *Cities and gender*. London: Routledge.

Lenclos, M. (2002). *Inclusive design: Access to London transport*. London: Royal College of Art.

Little, J., Peake, L., & Richardson, P. (1988). *Women in cities: Gender and the urban environment*. London: Macmillan.

Lucas, K. (2012). Transport and social exclusion: Where are we now? *Transport Policy, 20,* 105–113.

Madariaga, I. S., & Roberts, M. (Eds.). (2013). *Fair shared cities: The impact of gender planning in Europe*. London: Ashgate.

Manley, S. (1998). Creating accessible environments. In C. Greed & M. Roberts (Eds.), *Introducing urban design* (pp. 153–167). Harlow: Longmans.

Massey, D. (1984). *Spatial divisions of labour: Social structures and the geography of production*. London: Macmillan.

Matrix. (1984). *Making space: Women and the man-made built environment*. London: Pluto.

Mouffe, C. (2005). *The return of the political*. London: Verso.

ODPM. (2003). *Sustainable communities: Building for the future*. London: Office of the Deputy Prime Minister.

ODPM. (2005). *Diversity and equality in planning: A good practice guide*. London: Office of the Deputy Prime Minister.

Oliver, K. (1988). Women's accessibility and transport policy in Britain. In S. Whatmore & J. Little (Eds.), *Gender and geography* (pp. 19–34). London: Association for Curriculum Development.

ONS. (2012). *Social trends*. London: Office of National Statistics.

Plymouth. (2001). *Gender audit of the local plan review 2001 for the City of Plymouth*. University of Plymouth, School of Architecture, in association with City of Plymouth Council, written by M. McKie and team.

Reeves, D. (2005). *Planning for diversity: Policy and planning for a world of difference*. London: Routledge.

Roberts, M. (1991). *Living in a man-made world*. London: Routledge.

RTPI. (2003). *Gender mainstreaming toolkit*. London: Royal Town Planning Institute, by D. Reeves, C. Greed & C. Sheridan (Eds.), see www.rtpi.org.uk for subsequent material on gender and planning.

Stimpson, C., Dixler, E., Nelson, M., & Yatrakis, K. (Eds.). (1981). *Women and the American city*. Chicago: University of Chicago Press.

TCPA. (2009). *Planning for accessible and sustainable transport*. London: Town and Country Planning Association.

Uteng, T. P., & Cresswell, T. (2008). *Gendered mobilities*. London: Ashgate.

Vallance, S., Perkins, H., & Dixon, J. (2011). What is social sustainability? A clarification of concepts. *Geoforum, 42,* 342–348.

WDS. (2005). *Cycling for women*. London: Women's Design Service.

Whatmore, S., & Little, J. (Eds.). (1988). *Gender and geography*. London: Association for Curriculum Development.

CHAPTER 3

Travel Choice Reframed: "Deep Distribution" and Gender in Urban Transport

Caren Levy

WHY GENDER AND THE RIGHT TO THE CITY IN URBAN TRANSPORT?

It is some 25 years since the first work on gender and transport was published, including my own contribution focusing on the underlying gender assumptions in transport planning (Levy 1991, 1992). Much of this early work aimed at making women visible in urban transport, by showing the different travel patterns of women and men. However, while some offered deeper understandings of why these differences emerge and why they are overlooked in transport planning, many of the contributions were "restricted to the realm of behavioural differences" (Law 1999: 571).

To date, the literature that addresses gender in transport issues in cities and urban areas of the global South continues to expand but is still relatively scarce (Venter et al. 2007). This mirrors the relatively small proportion of investment and research in transport in urban areas of the global

C. Levy (✉)
University College London, London, UK
e-mail: c.levy@ucl.ac.uk

© The Author(s) 2019
C. L. Scholten, T. Joelsson (eds.), *Integrating Gender into Transport Planning*, https://doi.org/10.1007/978-3-030-05042-9_3

South, and the even smaller proportion of investment and research focused on addressing poverty and urban transport there (see Salon and Gulyani 2010: 642; also Srinivasan and Rogers 2005). Reflecting on the effectiveness of our understanding of the "transport–development interface", it is sobering to consider that

> The theory linking transport influences to social and economic change has not really been refined much beyond the general and aggregative levels. Few studies have addressed the matter of the distributional consequences of change nor derived comprehensive explanation to deal with this set of issues. (Leinbach 2000: 2)

Perhaps more concerning is that 25 years on, gender and all issues relating to distributional questions in transport are identified with "the social", which in turn is marginalized from the mainstream focus on economics, and now the burgeoning area of the environmental, in transport studies and planning (Jones and Lucas 2012).

With more than half the world's population living in cities, and with the predicted rates of urbanization and/or urban growth in Africa, Asia, Latin America and the Middle East, urban transport[1] is clearly a critical sector on which to focus current and future work. Moreover, changing urban structures and the resultant spatial distribution of activities and people in most medium and large-sized cities in the global South make transport an essential means for women and men, girls and boys to access activities in the city, enabling them to live lives of quality. Yet in most of these cities, "transport systems ... propagate an unfair distribution of accessibility and reproduce safety and environmental inequities" (Vasconcellos 2001: 5). Growing inequalities in cities, particularly under the impact of structural adjustment and globalization, have refocused attention on the right to the city (RTTC) of all its inhabitants (Harvey 2008; Marcuse 2010). The "unfair distribution of accessibility" in contemporary cities is a major constraint on particular groups of urban inhabitants being able to exercise their "choices" associated with the RTTC. In this context, Lefebvre's notions of the right to appropriation and the right to participation are useful concepts in examining gender in transport and "travel choice" (Lefebvre 1996).

This chapter will explore this relationship by deconstructing the notion of "travel choice" and reframing it in the context of the following: first, power relations and their intersectionality played out in private and public space; and second, the RTTC debates. By incorporating the RTTC, this chapter takes a normative approach, but it also argues for "the importance of the gender dimension in the construction of a **critical** understanding of our spatial concepts and our interpretation of the spatial behaviour of populations" (Cattan 2008: 83) in a wider understanding of urban development processes and practices.

REFRAMING "TRAVEL CHOICE"

The notion of "travel choice" is central to the modelling processes in mainstream transport planning. These models are based on

> "the paradigm of rational man", [sic] underpinned by "... neoclassical economic concepts, focusing upon the representation of people as individual rational choice makers, interacting together to form a state of equilibrium" (Avineri 2012: 513) and acting "... to maximize her utilities...applied to traveller behaviour to stimulate choices of destination, mode, route and time." (Avineri 2012: 523)

This conceptualization of travel choice is problematic, because it is based on an implicit assumption that "individual decision-making is made in a social vacuum" (Avineri 2012: 518). It does not recognize at least three critical issues central to transport and transport planning, namely the different social positions and multiple identities of transport users; the social construction of space, public and private; and the politics of transport in the context of social relations.

Social Position and Multiple Identities: Access and Control

Given the intersection of social relations, the social position of transport users reflects multiple identities of gender, class, ethnicity, religion, sexuality, age and mental/physical ability, which account for difference and inequality but which are also dynamic and open to change. To date, development planning of all kinds has not succeeded in conceptualizing, formulating, implementing and institutionalizing planning that addresses these multiple and simultaneous identities of women and men, (Levy 2009) and transport planning is no exception. The different and unequal social position of transport users reframes the traditional notion of travel choice that apparently maximizes utility. Rather, by virtue of the differently valued social roles and unequal access to and control over resources that different social positions offer, decisions about transport are a series of trade-offs about the purposes of travel, when to travel, how to travel and whether to travel or not. Moreover, these are not individual decisions but are taken in the context of unequal power relations in the household, in the community and in the city.

However, while mainstream transport planning does not recognize the social position of transport users, it is not ideologically neutral either. As the initial gender critiques of transport planning revealed, transport policy and planning is predicated on a number of implicit assumptions, which bias its outcomes. One such set of assumptions relates directly to gender relations and, in their understanding of the structure of households, the division of labour in households and the control of resources in decision-making in households, are informed by essentially "western" and middle-class values.[2] As many scholars and practitioners have argued, these assumptions are not based on any empirical reality in most urban contexts worldwide. Overlaid on these gender-based assumptions are other transport biases, namely a focus on the (male) journey to work and on motorized transport, particularly the private car, which is largely unaffordable to most poor urban women and men (Levy 1992). The overall outcome is a transport system that does not reflect the needs of the majority of urban dwellers, offering a range of transport options that are accessible only to some and thus does not provide the basis for making optimal travel choices. Neither do the travel patterns that emerge from these so-called travel choices reflect trips that were suppressed, re-routed or delayed (Avineri 2012; Vasconcellos 2001).

These debates about the distortions in travel choice clearly reflect questions of inequality, and therefore distributional issues, in mainstream transport planning. Recognizing the social position of transport users on the basis of their gender and its intersection with other social relations builds part of the foundation for understanding such distributional questions in transport. Other transport scholars[3] are also concerned with the significance of the social position of transport users through the lens of social exclusion, defined as

> the lack or denial of resources, rights, goods and services, and the inability to participate in the normal relationships and activities available to the majority of people in a society, whether in economic, social, cultural or political arenas. It affects both the quality of life of individuals and the equity and cohesion of society as a whole. (Levitas et al. 2007 in Lucas 2011: 1323)

The concept has been applied mainly in the UK, more recently in Europe, and only in a limited way in cities of the global South.[4]

The conflation and marginalization of distributional issues with "the social" have also been noted, and an argument made to re-conceptualize distribution from a peripheral to a mainstream issue in transport planning as follows:

> It would be more useful to define [transport] impacts and then recognize that each potentially has an economic, environmental and social dimension – all of which may have distributional consequences. (Jones and Lucas 2012: 5)

In this formulation, the authors define distributional impacts as spatial, temporal and socio-demographics. Issues of gender, class, age, ethnicity and race are placed under the socioeconomic category but recognized as operating in all three. While this "proposal" is extremely pertinent and timely, it still appears to fall within a behavioural paradigm, which does not problematize the power relations embedded in it.

Other critiques of utility-based notions of travel choice include recognition of beliefs and attitudes, emotion and "bounded rationality", perceived losses and gains (prospect theory), feedback and social learning (Avineri 2012: 513–515). Although these lines of enquiry raise interesting questions, at base they do not address issues of power or social position of individual travellers. This is so even in the case where, in examining the possibilities for behaviour change, reference is made to social dilemmas coming from "conflicting individual and collective interests" (Avineri 2012: 516). Nevertheless, individual and conflicting interests are central to an understanding of travel choice placed in a wider critique of the social construction of urban space and its interaction with transport.

Social Construction of (Public and Private) Space and Interests: Access, Mobility and Autonomy

The relationship between transport and land use has long been a concern of transport planning, although many would argue that transport decisions have not always been based on an explicit integration of these urban processes. Moreover, the understanding of transport and land use has been underpinned by a consensus view of society and does not explain urban processes in the context of the powerful forces driving capitalist development. Gender research in transport has made a particularly interesting contribution to such an understanding of urban development:

> Gender differences in transport contributed to a larger theoretical project in feminist geography: the critique of urban land use structure in contemporary capitalism, of the spatial separation of production and reproduction, and of the cultural dichotomy of public and private space. (Law 1999: 569)

Without detailing the debates in this "larger project", the following points are important to this discussion. Feminist geographers, while accepting that there is a close relationship between urbanization and capitalism, critique the proposition that urbanization—and consequently land development and use—is predominantly a "class phenomenon", as characterized by authors such as Harvey. They provide a coherent account of the intersection of class and gender in urban processes, demonstrating how gender is central to the notion of reproduction and therefore production. This relates to the increasing separation of home and work under capitalist urbanization and the false dichotomy between the association of private space with reproduction, consumption and women, and public space with production, politics and men. It is also a critique of Castells' framing of "the urban question" as being concerned with the "reproduction of labour power" rather than, or as well as, production, but without a comprehensive gendering of the notion of "reproduction of labour power" (Bondi and Peake 1988: 21). Finally, along with other scholars, feminist geographers challenge the dichotomous gender characterization between reproduction and production, along with separation of the private and public spheres, insisting that "greater consideration...be given, as in the treatment of production and reproduction, to the interaction and interconnectivity of the two spheres" (Little et al. 1988: 17), both structurally and in the space and time of everyday lives.

In an exploration of reproduction, building on the work of Benería (1979) and Harris and Young (1981), I demonstrate how these dichotomous gender characterizations are produced and reproduced by the intersection of all social relations, including class, ethnicity, religion and age, in the ideologies that underpin societal attitudes, including in our scholarly and professional disciplines (Levy 2009). One of the implications, in particular the conflation of biological reproduction with social reproduction, results in "a range of expressions of men's control [including state control] of women's reproduction in nearly all societies", which includes "restrictions on women's mobility" (Levy 2009: v).

From the point of view of transport, three related factors emerge in reframing the notion of "travel choice" from this analysis of the urban. The first is that decisions about travel are not based on the social position of women and men alone. They are also influenced by the interrelationships between transport in urban areas and the "socioeconomic, political and environmental processes [that] create, reproduce and transform not only the places in which we live, but also the social relations between men and women in these places" (Little et al. 1988: 1–2). This accounts for the reproduction of powerful dominant interests not only in the transport system, but also in the spatial structure and land uses of the city, creating a framework of inequality in which decisions about travel are made. This is an understanding of transport and transport planning as conflictual and contested, a far cry from the consensual paradigms of mainstream transport planning.

The second factor is that the intrinsic "public space nature" of transport is not problematized at all in behavioural approaches to transport and travel patterns. Most women—and some men—do not have the full autonomy to move in public space because of the social control exercised in the interests of dominant gender, class, ethnic, religious, sexuality and age groups. Therefore, a third factor to consider in an understanding of "travel choice" is that decisions to travel are made on the basis of norms exercised in the private as well as the public sphere, based on support and/or permission, persuasion and/or prohibition, verbal and/or physical. It is no accident that along with providing a deeper understanding of gender roles, the primary focus of research on gender and transport has been on male sexual violence in public and private space, and the constraints on women's travel patterns to and from employment, with a respective focus on fear of violence and social relations in the household and the workplace "as central mechanisms of oppression" (Law 1999: 569).

Along with a recognition of the social position of transport users on the basis of gender and other intersecting social relations, understanding transport in the context of the "interrelations between socially constructed gender relations and socially constructed environments" (Little et al. 1988: 2) acknowledges the "deep distributional" issues in urban transport and urban transport planning. An approach that recognizes "deep distribution" builds the foundations for an understanding of transport based on the articulation of power relations in public and private space at the level of the household, community and society that generate the structural inequality and dominant relations under which decisions about "travel choice" are negotiated and made.

Scales of Gendered Spatial Politics: The Right to the City

Treating the notion of travel choice in a "social vacuum" results in an understanding of transport that is also devoid of politics. How can such a central urban investment with such widespread implications for urban inhabitants, from land values to forced evictions to access to the livelihood and social opportunities offered by the city, be devoid of politics? In reframing travel choice as a political act rather than a technical exercise, the contemporary discussion related to the "right to the city" (RTTC) is attractive.

These interrelationships between a biased transport system, the social and unequal position of transport users, and socially constructed inequality in urban environments take on a particular form in the restructuring of contemporary cities, where transport is a central sphere of action in what Harvey (2008) refers to as "processes of creative destruction". Although these processes will not be explored in detail in this chapter, their recognition is important to the context in which "travel choices" and the RTTC are framed. Deregulation and the penetration of the market into areas previously controlled and managed by the state have resulted in a reconfiguration of public and private relationships and space in the city. Investment in transport is a good reflection of these changes, as the modernization of transport infrastructure is seen as a critical driver in making cities globally competitive in the image of the "world class city". In cities all over the global South the result has been the "accumulation of dispossession", expressed in a number of impacts including market and forced evictions (see, e.g. Hasan 2006; also Patel et al. 2002) and a transport system, implemented through new partnerships between private service providers and the state, which is out of the reach of the majority of poor urban inhabitants. Cities are sites of "processes of creative destruction that have dispossessed the masses of any right to the city whatsoever" (Harvey 2008: 37).

Furthermore, "The right to the city is far more than the individual liberty to access urban resources: it is a right to change ourselves by changing the city" (Harvey 2008: 23). Interpreting this much-quoted line from Harvey from a gender perspective, the RTTC offers hope for change not only for our future cities and their governance but also for our "socially constructed" selves, so that women and men, girls and boys can live everyday lives in both recognition and equality. Current formulations of change in transport studies focus on the "design of 'social nudges' to change travel behaviour" (Avineri 2012: 518) "to make choices that are better for them and society" (Avineri 2012: 516, writing in the context of environment and climate change). This formulation of change is problematic in two senses. It runs the risk of directing attention away from structural inequalities and issues of "deep distribution", making individuals "the problem". Equally important, the question of consciousness is also absent. Claiming the RTTC requires a change in individual and collective consciousness, challenging prevailing ideologies through an awareness and a rejection of current structural inequalities, and the collective construction of a mobilizing vision of a more socially just city in which equal opportunity for material well-being, recognition and democratic decision-making are central principles. What can transport offer this transformative process of consciousness raising? In the following sections, it is argued that the contested intersection of the public and private spheres from which decisions relating to transport are negotiated offers some interesting potential for changing individual and collective consciousness.

Finally, in the context of the dominant apolitical character of transport planning, the RTTC is a clear statement of the opposite. "Lefebvre's right is both a cry and a demand, a cry out of necessity and a demand for something more" (Marcuse 2010: 190). This firmly roots transport, and the individual and collective choices made about it, in a political framework. It also roots it in a discussion of urban citizenship, which "is not so much about legality than about legitimacy" (Zérah et al. 2011: 4). Urban citizenship is about the right to be in and of the city. In cities of the global South, for women and men excluded in different context-specific ways from the city on the basis of their gender and other intersecting social identities, such legitimacy is fundamental and often the basis for moving beyond either being invisible or being bulldozed to more constructive policies and planning.

However, using the banner of RTTC for transport will not automatically lead to a supportive policy environment for the deprived, dominated and oppressed in the city. This is an ongoing struggle, the dimensions of which need to be operationalized, in this case from an urban transport perspective. Lefebvre argues that "the right...to participation and appropriation...are implied in the right to the city" (Lefebvre 1996: 174). The following is a discussion of how each might have theoretical and practical application to an inclusive and democratically formulated, implemented and managed transport system in the city.

The right to appropriate implies the use and occupation of urban space. This is clearly differentiated from the notion of domination and exchange, (Lefebvre 1991) which would encompass property ownership and the increasing privatization of space in the contemporary city. Transport is unique in its capacity to engage with the right to appropriation in two ways. First, the access provided by transport is one dimension of the right to appropriate urban space for a range of uses that are important to the daily lives of different and unequally positioned women and men, girls and boys. Transport access is understood not only in geographical terms as the distance between different locations in the city involving movement at different levels of public space. It also encompasses economic concerns related to affordability, sociocultural aspects related to safety and security in public space, and the physical issues of comfort and ease of design in the use of transport and its related infrastructure. Second, the act of travelling is a form of appropriation of public space in the city in its own right. The freedom to move in public space without physical or verbal threat is both of material and symbolic meaning to women and to particular men (e.g. on the basis of racial or sexual identities) in ways that are context-specific.

While the first form of appropriation relates to accessibility, the second form could be said to relate to the question of mobility. The two work together in understanding transport and the right to appropriation, based on a particular understanding of mobility. Historically, the distinction between the two and an emphasis on accessibility has been at the core of more progressive views of transport (Levy 1992; also Vasconcellos 2001), well highlighted in the following quotation about the current emphasis on mobility as an end in itself in Indian urban transport planning:

> The solutions offered by transport policies and public–private partnerships posit methods of transit that purport to be universally accessible, yet are in empirical fact out of the reach of the majority... Rather than ensuring the directness of links and a density of connectors, by accounting for geographical destination of activities, mobility in and of itself becomes a performance objective in planning, with greater public investment in roads easily accessible by car users rather than modes that facilitate multi-modal access for heterogeneous publics. The unresolved dilemma in the Indian planning scenario is on where good mobility is seen as a sufficient condition for accessibility. (Murthy 2011: 122)

However, current research provides a new focus on "mobilities" as a way to view contemporary processes and experiences of globalization and technological change, either "positively coded as progress, freedom or modernity itself" or raising "issues of restricted movement, vigilance and control" (Cresswell and Uteng 2008: 1). There is also a current tendency to use "mobility" instead of transport "to signal the new framing of the topic within social and cultural geographies of mobility" (Law 1999: 568). The right to in public space is just such an issue and a way to challenge, theoretically and in practice, the social control exerted on different groups to move freely in public space in the name of identity-based ideologies presented as "natural" or in some cases "god given".

Alongside the right to appropriation, the right to participation is a key dimension of the RTTC's focus on "the exercise of a collective power to reshape the processes of urbanization" (Harvey 2008: 23). Thus, the right to participation demands

> greater democratic control over the production and utilization of the surplus. Since the urban process is a major channel of surplus use, establishing democratic management over its urban deployment constitutes the right to the city. (Harvey 2008: 37)

Keeping in mind Lefebvre's insistence on a "real and active participation" (Lefebvre 1996: 145), transport also offers a two-fold view of participation. First, it focuses on the right to participate in decision-making about urban transport, something that has been limited so far in transport planning in most parts of the world. Second, participation also encompasses the access provided by transport to travel in order to participate in decisions about the making of the built environment and the politics of the city. Both kinds of participation involve an engagement with the public sphere of politics, traditionally the realm of men and one in which women face many constraints related to gender and its intersection with other social relations.

GENDER RELATIONS IN TRANSPORT: THE RIGHT TO APPROPRIATION

From the research undertaken over the last 25 years, what can be said about transport and how it relates to the two-fold understanding of the right to appropriation presented above—that is, the access provided by transport to enable the appropriation of urban space for a range of uses that are important to the daily lives of women and men, girls and boys— and the capacity to exercise autonomous agency around being mobile in public space?

Some clear patterns have emerged in the research on how gender impacts on transport. However, one needs to be careful about what it tells us about the right to appropriate the city. For a start, they reflect an aggregation of individual travel, not a collective use and occupation of space. Nevertheless, transport patterns disaggregated by gender do indicate that women and men use the city differently in terms of purpose, location, time and mode of travel. Some of the research shows the impact of multiple identities on the use of transport, with patterns correlated with differences in income, education and location in the city (Srinivasan and Rogers 2005; Venter et al. 2007) as well as with marital status and household composition.

Thus, in most contexts men tend to travel more than women. While work trips account for the highest proportion of both women and men's trips, men tend to undertake more of them and to more distant locations— see, for example, Hanoi, Vietnam (Fig. 3.1). Women tend to do more trips related to reproductive work than men and more multi-purpose trips, or "trip chaining". There are also clear temporal gender differences. In most contexts, men do more peak travelling than women, when service provision is relatively better, and women tend to do more off-peak trips during the day than men, when transport services are less frequent—see, for example, Lima, Peru (Fig. 3.2).

Women tend to use different and cheaper modes of transport than men, particularly low-income women—for example, the modal use in Johannesburg, South Africa, which shows the intersection of gender, race and, implicitly through the mode, class (Fig. 3.3). Giving some clues to

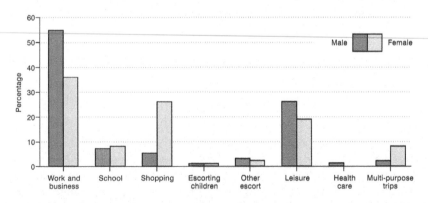

Fig. 3.1 Purpose of trip according to gender, Hanoi, Vietna. (Source: Adapted from Tran and Schlyter 2010)

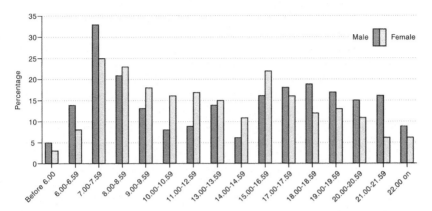

Fig. 3.2 The use of public transport by gender and time, Lima, Peru (multiple responses). (Source: Adapted from Gómes 1997: 11)

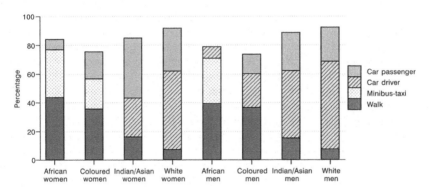

Fig. 3.3 Modal choice by gender and race, Johannesburg, South Africa. (Source: Adapted from 2002 census data for Johannesburg cited in GTZ 2007: 7)

inter-household distribution of resources and negotiation, where there is a family car, men will tend to use it, also keeping in mind that in cities in the global South women are less likely to have driving licences than men. Finally, poor women and men tend to walk more than those who are better off, with poor women tending to walk most (see, e.g. Anand and Tiwari 2006; also Salon and Gulyani 2010). The lack of "travel choice", resulting in a trade-off between cost and modal selection, is clearly expressed in the context of Nairobi slums, which "are relatively well connected by privately owned and operated transit vans and small buses called matatus. The slum residents are walking largely because they cannot afford the motorized options." (See Salon and Gulyani 2010: 641–642.) Accidents are a serious problem for pedestrians, who often use roads with no pavements and in conflict with road traffic. However, the death rate for men killed on roads tends to be higher than for women. For example, in São Paulo, men account for 76 per cent of pedestrian fatalities and 86 per cent of vehicle fatalities (Vasconcellos 2001).

Not unsurprisingly, the issue of safety and security in the different public spaces of transport modes (e.g. buses and taxis), routes (e.g. streets) and hubs (e.g. bus stations) is widely addressed. Women tend to experience more violence through thefts and verbal and physical/sexual abuse while waiting for and using public transport. Some groups of men also experience abuse on transport, for example, around their race or sexuality. Violence is a problem for women on all transport modes and routes. For example, a study in Delhi, India, found that

> Women are the targets of sexual harassment while travelling to work and practically every woman interviewed had anecdotal evidence of suffering from the same. Harassment while walking down the street or travelling on a bus is a common occurrence for working women and is exacerbated by the absence of adequate lighting on streets and subways and by the small, lonely paths connecting the slum with the bus stops. (Anand and Tiwari 2006: 78)

During the Arab Spring and in its aftermath in Cairo, Egypt, women experienced and are still experiencing high levels of sexual harassment in public spaces, and the following statement encapsulates the challenge of women's right to appropriation vividly:

> despite claims made by men, harassment is not a harmless, direct reaction by men to women but an institutionalized system of violence that functions to police women's participation, freedom of movement and behaviour in public spaces. It is not how women behave in the public sphere that makes them vulnerable to street harassment; it is that they have chosen to enter the public sphere at all. (El Nahry 2012)

A critical caveat in these general findings is that political, socioeconomic and environmental context is all important, and differently located cities in time and space will certainly account for very different decisions about, and experiences of, travel by different groups in different parts of different cities (Hanson 2010). This research is also mainly quantitative, and the disaggregation of travel patterns by sex gives a clear view of the gender division of labour. As the research acknowledges, these patterns reflect that while in most societies women retain domestic and childcare responsibilities, increasing numbers of women are balancing these with productive work. Indeed, there is a concentration of research focused on travel to work that

> derives in part from its power as a metaphor... it is the single human activity that most clearly bridges the symbolic and spatial distinction between public and private which is a feature of western urbanism. It is the actual and metaphoric link between the spheres and spaces of production and reproduction, work and home. (Law 1999: 571)

Furthermore, travel to work is a reflection not only of the need and desire of women to work but also of the spatial use of the city and its intersection with the gendered spatial labour markets in cities (see, e.g. Hanson and Pratt 1995).

However, we do not have a deeper understanding of the trade-offs diverse women and men have made in their decisions to travel, which requires complementary qualitative research. With few exceptions, (see, e.g. Marome 2009), the research does not reflect the negotiation within the private sphere of the household about women and men's travel in the public sphere. What are the possibilities for changing gender relations in either sphere because of the dynamic relationship between decisions in the private sphere and accessibility and mobility in public space?

This challenges two related dimensions of the RTTC discourse. First, it points to the importance of recognizing that the relationship between the appropriation of space in the public sphere is contingent on and influenced by decisions in the private sphere. The decision to travel is negotiated in the household because it implies moving into the public space of the city. Focusing on the question of consciousness, scholarship outside the transport field suggests that women's negotiation in the household is based on positions that derive neither from false consciousness nor from conscious collusion but from positions of power available to them (Kandiyoti 1988). At the same time, women's appropriation of the city through mobility in public space and accessibility to what the city offers may strengthen women's agency and autonomy, with implications for relations in the household. We do have some insights into these issues but they generate more questions than answers.

Second, it highlights how the right of appropriation through equal transport access to necessities and opportunities in the city is both conceptual and practical. With the separation of home and work and the growth, dispersal and restructuring of cities, transport is a use value that has become a necessity for urban inhabitants. "Only state intervention can achieve the levels of security, social welfare and (relative) equity of access to mobility" (Docherty et al. 2004: 259).[5] However, transport is commodified in most cities, and with the advent of neo-liberal policies, private operators driven by profit are increasingly involved in urban transport provision, which is increasingly expensive to access. This shifts the discussion from the use of transport to facilitate the right to appropriate the city, to rights in the city (Parnell and Pieterse 2010; Zérah et al. 2011), for example, the right to employment, the right to housing and the right to affordable, convenient and safe transport.

GENDER RELATIONS IN TRANSPORT: THE RIGHT TO PARTICIPATION

In most contemporary cities, the right to participation in decision-making about transport is a demand in the form of collective protest against transport planning decisions already taken. These are usually angry collective responses to either the way transport systems are designed or managed—for example, mobilization around higher fares—or to proposals or forced evictions in order to build new or extend old infrastructure. Perhaps more accurately defined as appropriation in the name of recognition and participation, the latter is an increasingly common event in the context of contemporary urban restructuring in which the upgrading of transport infrastructure is a central strategy. For example, the upgrading of transport in Mumbai and Delhi, India, is seen as a key intervention in the creation of "world class cities" (Patel et al. 2002; Anand and Tiwari 2006). In the context of the World Bank–supported Mumbai Urban Transport Plan, despite ongoing negotiations by organized communities with Indian Railways and other government agencies, slum dwellers along the railway line were bulldozed. The National Slum Dwellers Federation, which incorporates Mahila Milan, a women's collective, occupied the Kanjur Marg railway line in protest, bringing services to a halt in a show of strength to bring Indian Railways and its local partners back to the negotiating table (Burra 1999).

Such forms of temporary appropriation are critical moments in an ongoing struggle to demand more participation and democratic governance in the planning of urban transport. As a largely expert-led and top-down intervention, there are not many positive examples worldwide of active support for public participation in transport planning, let alone where this has been done in a gender-sensitive manner. Indeed

> to develop an effective transport policy framework which embraces bottom-up strategies will require a sea change in the traditional attitudes of transport experts and the organizational culture of the profession. (Booth and Richardson 2001: 148)

Although this is a statement made in the context of the UK, it applies worldwide. The current discourses about participation in transport planning, with the exception of those concerned with social exclusion, (Hodgson and Turner 2003), are either "extractive" in character, focusing on wider consultation with travellers to improve data in transport surveys, or relate to "partnership structures...and mean more about partnership of a wider range of professional service providers than users" (Hodgson and Turner 2003: 269).

An understanding of gendered access and mobilities could also contribute to more hybrid notions of collective power. Politics in urban areas is imbued with a spatiality, either at the point of production (trade unions, workers' organizations) or at the point of residence (community-based organizations organizing over issues of "collective consumptions"). Often, this is presented as a dichotomy of male and female realms, respectively, which belies the interaction and fluidity between them. At the point of production, transport can be a focus of mobilization, for example the safety of women workers when travelling to and from work (see, e.g. on women IT workers in Mexico City, Ruiz Castro 2009) or a means to get safely to workers meetings and demonstrations. At the point of residence, transport may be the focus of residentially based political organization, a scale at which Moser (1987) highlights the gendered difference between the community management roles of women (organizing around issues related to reproduction at the point of residence) and the community politics roles of men, the latter often carrying more access to resources and power.

Because transport implies accessibility and mobility to citywide activities and opportunities, it also offers a cross-scalar configuration of politics, enabling women and men to access different political spaces in the city. Debates about extending public participation in transport planning in the UK recognize this more strategic entry point for participation at the city-wide scale as well as in more local transport planning (Booth and Richardson 2001; Hodgson and Turner 2003). This opens up interesting but complex spaces for the right to participation. Experience in other areas of planning would suggest that women are more likely to be included in participatory initiatives at the local level, and explicit steps would have to be taken to connect to existing gender-based movements to ensure their equal participation at city level. Women's almost universal under-representation in formal (party) politics "must not be confused with absence of any political activity" (Bondi and Peake 1988: 21) since women are often also involved in more "informal" politics at community level.

The role of transport in facilitating the right to participation in other policy and planning in the city is under-researched. What is clear is that the right to participate, whether in transport or other urban planning, is a struggle that requires the construction of a collective consciousness and social movements around the RTTC.

> Thus, strategies for a progressive transformation lie in a complex process of political alliance building, on the basis of intersecting identities where common sources of exclusion, exploitation and oppression are acknowledged, and the interlinked agendas for recognition and redistribution are brought together. (Levy 2009: viii)

Possible forms for such movements are various—a forum, a coalition, an alliance, a movement, an assembly (Marcuse 2010: 192)—but the challenge in constructing them is great. The very difference in social positions of urban dwellers and their translation into conflictive interests in the city, which have been highlighted in this chapter as critical to a deeper under-standing of decisions to travel, are precisely those forces that divide political movements. For example, with respect to class differences in Chennai:

> the right to the city as a concept is highly ambiguous, given that there are multiple understandings of open spaces and what the right to the city might mean. Middle-class activism does not take account of these multiple under-standings and, as a result, the right to the city becomes subservient to bourgeois citizenship. For the poor, there is no right to the city. (Arabindoo 2010: 10)

Tactics are also important and should not be conflated with the discourse of "rights talk". Thus, while women in Shack Dwellers International and the Asian Coalition of Housing Rights recognize their experience of inequality and injustice in the city, they "seek the 'identity' of constructive engagement, not of insurgency" (Mitlin 2010: 8), which they associate with rights.

Nevertheless, transport is an issue that has the power to cut across difference. For example, the question of safety and security along transport routes and in transport modes alone is an issue that can mobilize a city. While the current discourse on the RTTC focuses on collective rights, when it comes down to it, claiming the collective right to participate is contingent on decisions in the household as well as in the local and wider community, reflecting an interaction between the private and the public realm about which we need more qualitative research. Does access to urban activities and mobility in public space empower (or disempower) different women and men? In what ways? How do they take this experience back into decision-making in the household? How do they take it into urban politics? Responding to these kinds of questions is crucial to deepening an understanding of multi-scalar social relations in urban contexts. "Being mobile is not just about geographical space but also, and probably above all, about social space" (Cattan 2008: 86), and political space as the right to participate in the struggle for the RTTC.

Transport Planning Revisited

This chapter argues that a paradigm shift in transport planning is way overdue. Some 25 years after researchers and practitioners began to highlight the importance of gender relations in urban transport, there is still a lacuna in transport planning's understanding of the everyday lives of women and men, girls and boys in urban areas. Exploring the "deep distributional" dimensions of "travel choice" in urban transport and transport planning reframes not just the understanding of social relations and urban transport but also the very behavioural foundations of transport planning itself.

However, reframing understandings of diversity and inequality

> in paradigms which are deeply embedded in policy making and planning involves more than the power of rational argument. It is closely intertwined with political processes and a change in consciousness which questions and deconstructs the fundamentals of current power relations and their expression in diverse and multiple identities in contemporary societies. (Levy 2009: vi)

For this reason, the mobilizing potential of the right to the city may be a useful contribution to the struggle for change in transport planning. "Understanding the urban form in development projects and transportation needs to recast the understanding of the public as a space of common interest, a collective aspiration", Murthy argues in the context of India (Murthy 2011: 131).

However, as stated at the beginning, the arguments developed in this chapter are not only normative. A gender and RTTC perspective strengthens not only a critical understanding of transport in urban development but also the predictive capacities of transport planning and its capacity to respond to urgent urban development challenges in the context of active democratic politics.

If "public space must be understood as a gauge of the regimes of justice extant at any particular moment" (Mitchell 2003: 235 following Van Deusen 2002), where does this leave the current state of gender relations in transport and women's access to public space in our cities and urban areas?

NOTES

1. In this chapter, urban transport is understood as the interrelationship between transport and urban processes, and the practices of transport planning—which together could be understood as a transport system.
2. On assumptions in policy and planning in general, see Moser (1989). On assumptions in transport planning, see Levy (1992).
3. See, for example, Preston and Rajé (2007); also Stanley and Lucas (2008); and Stanley and Vella-Brodrick (2009).
4. For example, for an exploration of social exclusion and "transport poverty", see Lucas (2011).
5. Their argument is that the state intervenes in transport, seeking to maximize the social value of transport as a public good, following Schumpter's notion of a public good.

REFERENCES

Anand, A., & Tiwari, G. (2006). A gendered perspective of the shelter–transport–livelihood link: The case of poor women in Delhi. *Transport Reviews, 26*(1), 63–80.

Arabindoo, P. (2010). In T. Bastia, M. Lombard, H. Jabeen, G. Sou and N. Banks with inputs from the contributors and workshop convenors, C. O. N. Moser, & M. Herbert, *Right to the city workshop report*. Manchester: Urban Rights Group, University of Manchester.

Avineri, E. (2012). On the use and potential of behavioural economics from the perspective of transport and climate change. *Journal of Transport Geography, 24,* 512–521.

Benería, L. (1979). Reproduction, production and the sexual division of labour. *Cambridge Journal of Economics, 3,* 203–225.

Bondi, L., & Peake, L. (1988). Gender and the city: Urban politics revisited. In J. Little, L. Peake, & P. Richardson (Eds.), *Women in cities: Gender and the urban environment* (pp. 21–40). London: Macmillan.

Booth, C., & Richardson, T. (2001). Placing the public in integrated transport planning. *Transport Policy, 8,* 141–149.

Burra, S. (1999). *Resettlement and rehabilitation of the urban poor: The story of Kanjur Marg.* DPU Working Paper No 99, DPU, University College London.

Cattan, N. (2008). Gendering mobilities: Insights into the construction of spatial concepts. In T. P. Uteng & T. Cresswell (Eds.), *Gendered mobilities.* Aldershot: Ashgate.

Cresswell, T., & Uteng, T. P. (2008). Gendered mobilities: Towards an holistic understanding. In T. P. Uteng & T. Cresswell (Eds.), *Gendered mobilities.* Aldershot: Ashgate.

Docherty, I., Shaw, J., & Gather, M. (2004). State intervention in contemporary transport. *Journal of Transport Geography, 12*(4), 257–264.

El Nahry, F. (2012). She's not asking for it: Street harassment and women in public spaces. *Gender across borders: A global voice for gender justice.* Available at http://www.genderacrossborders.com/2012/03/20/shes-not-asking-for-it-streetharassment-and-women-in-public-spaces/

Gómes, L. (1997). "Schedules for Lima public transportation", cited in GTZ (2007), *Gender and urban transport: Smart and affordable. Sustainable transport: A sourcebook for policy makers in developing cities.* Available at http://www.itdp.org/documents/7aGenderUT%28Sept%29300.pdf. 141–149.

GTZ. (2007). Gender and urban transport: Smart and affordable. *Sustainable transport: A sourcebook for policy makers in developing cities.* Available at http://www.itdp.org/documents/7aGenderUT%28Sept%29300.pdf

Hanson, S. (2010). Gender and mobility: New approaches for informing sustainability. *Gender, Place and Culture, 17*(1), 5–23.

Hanson, S., & Pratt, G. (1995). *Gender, work and space.* New York: Routledge.

Harris, O., & Young, K. (1981). Engendered structures: Some problems in the analysis of reproduction. In J. S. Kahn & J. R. Llobera (Eds.), *The anthropology of pre-capitalist societies.* London: Macmillan.

Harvey, D. (2008). The right to the city. *New Left Review, 53,* 23–40.

Hasan, A. (2006). *Livelihood substitution: The case of the Lyari Expressway.* Karachi: Ushba International Publishing.

Hodgson, F. C., & Turner, J. (2003). Participation not consumption: The need for new participatory practices to address transport and social exclusion. *Transport Policy, 10,* 265–272.

Jones, P., & Lucas, K. (2012). The social consequences of transport decision-making: Clarifying concepts, synthesizing knowledge and assessing implications. *Journal of Transport Geography, 21,* 4–16.

Kandiyoti, D. (1988). Bargaining with patriarchy. *Gender and Society, 2*(3), 274–290.

Law, R. (1999). Beyond 'women and transport': Towards new geographies of gender and daily mobility. *Progress in Human Geography, 23*(4), 567–588.

Lefebvre, H. (1991). *The production of space.* Oxford/Cambridge, MA: Blackwell Publishers.

Lefebvre, H. (1996). *Writings on cities.* Oxford: Blackwell Publishers.

Leinbach, T. R. (2000). Mobility in development context: Changing perspectives, new interpretations and the real issues. *Journal of Transport Geography, 8*(1), 1–9.

Levitas, R., Pantazis, C., Fahmy, E., Gordon, D., Lloyd, E., & Patsios, D. (2007). *The multi-dimensional analysis of social exclusion.* Project report. University of Bristol.

Levy, C. (1991). *Towards gender aware urban transport planning, gender and Third World development training package, module 5.* Brighton: IDS Publications.

Levy, C. (1992). Transport. In L. Ostergaard (Ed.), *Gender and development,* Chapter 6 (pp. 94–109). London: Routledge.

Levy, C. (2009). Gender justice in a diversity approach to development? The challenges for development planning, (viewpoint). *International Development Planning Review, 31*(4), i–xi.

Little, J., Peake, L., & Richardson, P. (1988). *Women in cities: Gender and the urban environment.* London: Macmillan Education.

Lucas, K. (2011). Making the connections between transport disadvantage and social exclusion of low-income populations in the Tshwane region of South Africa. *Journal of Transport Geography, 19,* 1320–1334.

Marcuse, P. (2010). From critical urban theory to the right to the city. *City, 13*(2–3), 185–197.

Marome, W. (2009). *A gendered spatial analysis of the relationship between women's productive work and women's autonomy in the household: Understanding women's agency in public and private space in a Bangkok soi.* PhD thesis, University College London.

Mitchell, D. (2003). *The right to the city: Social justice and the fight for public space.* New York/London: The Guilford Press.

Mitlin, D. (2010). In T. Bastia, M. Lombard, H. Jabeen, G. Sou and N. Banks, with inputs from the contributors and workshop convenors, C. O. N. Moser, & M. Herbert, *Right to the city workshop report.* Manchester: Urban Rights Group, University of Manchester.

Moser, C. O. N. (1987). Mobilization is women's work: Struggles for infrastructure in Guayaquil, Ecuador. In C. Moser & L. Peake (Eds.), *Women, human settlements and housing* (pp. 166–194). London: Tavistock.

Moser, C. O. N. (1989). Gender planning in the Third World: Meeting practical and strategic gender needs. *World Development, 17*(2), 1799–1825.

Murthy, K. (2011). Urban transport and the right to the city: Accessibility and mobility. In M.-H. Zérah, V. Dupont, & S. Tawa Lama-Rewal (Eds.), *Urban policies and the right to the city in India: Rights, responsibilities and citizenship* (pp. 122–132). New Delhi: UNESCO/CSH.

Parnell, S., & Pieterse, E. (2010). The right to the city: Institutional imperatives of the development state. *International Journal of Urban and Regional Research, 34*(1), 146–162.

Patel, S., d'Cruz, C., & Burra, S. (2002). Beyond evictions in a global city: People managed settlement in Mumbai. *Environment and Urbanization, 14*(1), 159–172.

Preston, J., & Rajé, F. (2007). Accessibility, mobility and transport-related social exclusion. *Journal of Transport Geography, 15*(3), 151–160.

Ruiz Castro, M. F. (2009). Empowerment and gender in the workplace: Experiences in accounting and IT firms in Mexico. PhD thesis, University College London, 310 p.

Salon, D., & Gulyani, S. (2010). Mobility, poverty and gender: 'Travel choices' of slum residents in Nairobi, Kenya. *Transport Reviews, 30*(5), 641–657.

Srinivasan, S., & Rogers, P. (2005). Travel behaviour of low-income residents: Studying two contrasting locations in the city of Chennai, India. *Journal of Transport Geography, 13*, 265–274.

Stanley, J., and Lucas, K., 2008. Social exclusion: What can public transport offer?. Research in Transportation Economics, 22(1), pp. 36–40.

Stanley, J., & Vella-Brodrick, D. (2009). The usefulness of social exclusion to inform social policy in transport. *Transport Policy, 16*, 90–96.

Tran, H. A., & Schlyter, A. (2010). Gender and class in urban transport: The cases of Xian and Hanoi. *Environment and Urbanization, 22*(1), 139–155.

Van Deusen, R. (2002, June 1). *Urban design and the production of space in Syracuse, New York.* Paper presented at the right to the city conference, Rome.

Vasconcellos, E. (2001). *Urban transport: Environment and equity: The case for developing countries.* London/Sterling: Earthscan.

Venter, C., Vokolkova, V., & Michalek, J. (2007). Gender, residential location and household travel: Empirical findings from low-income urban settlements in Durban, South Africa. *Transport Reviews, 27*(6), 653–677.

Zérah, M.-H., Tawa Lama-Rewal, S., Dupont, V., & Chaudhuri, B. (2011). Introduction: The right to the city and urban citizenship in the Indian context. In M.-H. Zérah, V. Dupont, & S. Tawa Lama-Rewal (Eds.), *Urban policies and the right to the city in India: Rights, responsibilities and citizenship* (pp. 1–11). New Delhi: UNESCO/CSH.

PART II

Instruments for Change

Gendered Perspectives on Swedish Transport Policy-Making: An Issue for Gendered Sustainability Too

Lena Smidfelt Rosqvist

INTRODUCTION

Gendered perspectives on mobility deserve attention because mobility is a fundamental part of everyday life which looks different for different groups, such as men and women, in society. Research has problematized gender equality in the transport sector and made efforts to make women visible in transport, mainly by focusing on differences in mobility (e.g. Levy 2013; Polk 1998). While gender equality matters everywhere, in the transport sector it has a justification of its own as decisions made for the transport sector affect women's and men's individual mobility, and thus their lives, by such means as favouring different modes of transport.

Climate change is probably the greatest challenge of our time, and the transport sector plays an important role in reaching climate objectives due to the reduced carbon dioxide (CO_2) emissions in sustainable transport strategies (May 2013). Measures to achieve this goal require a move away from single-occupancy, fossil-fuelled vehicles through a combination of

L. Smidfelt Rosqvist (✉)
Trivector Traffic AB, Lund, Sweden
e-mail: lena.smidfelt@trivector.se

© The Author(s) 2019
C. L. Scholten, T. Joelsson (eds.), *Integrating Gender into Transport Planning*, https://doi.org/10.1007/978-3-030-05042-9_4

69

the development and implementation of new solutions and changes in people's everyday travel behaviour (Nilsson et al. 2013; Commission on Fossil Free Road Transport 2013; Moriarty and Honnery 2013). The need for a reduction in the total distance driven by private cars (in addition to converting the vehicle fleet and fuel) has been estimated for Sweden as 12% by 2030 and 18% by 2050 compared to 2010 (Swedish Transport Administration 2012a: 224: 37). In Sweden, as well as globally, great efforts have been put into finding solutions and measures to reduce the environmental impact of the transport sector (e.g. Commission on Fossil Free Road Transport 2013; European Commission 2011). Even so, transport is still increasing its share of emissions (Swedish Environmental Protection Agency 2015; European Environment Agency 2017). When transport behaviour is analysed from the perspective of gender, women's transport behaviour in general terms is more in line with what is required for a transition that favours climate and sustainability objectives, due to men travelling greater distances and more by car (e.g. Polk 2003; Kronsell et al. 2015).

Based on the understanding that transport behaviour on average needs to change if climate targets for the transport sector are to be reached, climate-sustainable patterns of transport could be modelled on women's transport choices, as discussed by Kronsell et al. (2015). Gender equality for the transport sector thus *also*—apart from reasons of democracy—plays a role in climate sustainability targets. Moreover, women tend to have stronger *preferences* for measures that will improve overall sustainability in the transport sector, and they are also more willing to change or act on climate concerns (Swedish Environmental Protection Agency 2009; Kronsell et al. 2015). To me, this emphasizes yet another gender equality issue. Average preferences by women, as defined by Dymén et al. (2013) using a categorization into masculine and feminine attributes in planning based on Kurian (2000), such as concern for the environment and political rationality, have in fact been identified as likely to bring the issue of climate change into planning (Dymén et al. 2013). The climate change challenge thus provides a strong argument that a gendered perspective should be included in transport policy beyond the ambition to increase the democratic quality of policy-making.

The point of departure for this chapter is my experience as a professional transport planner, researcher and national expert on sustainable transport. In my work—analysing what is needed for the transport sector to achieve policy sustainability targets—I have long been aware of the discrepancy between what targets stipulate and the decisions that are made (e.g. Smidfelt Rosqvist and Wennberg 2012; Smidfelt Rosqvist 2009). With a specific interest in travel patterns, as well as analyses of measures with the power to achieve sustainability, I have long been aware of the possible fit between gendering the transport sector and achieving climate sustainability for this sector (Kronsell et al. 2015). Furthermore, between 2010 and 2015, I acted as chair of *Nätverket för kvinnor i transportpolitiken*, an NGO with the aim of contributing to achieving the target of Swedish national policy on gender equality in the transport sector. This position provided an opportunity to initiate a couple of investigations into gender equality in the Swedish transport sector.

In this chapter, I will draw on my vast experience of working as a senior expert in transport planning, using gendered statistics provided by the Statistics Sweden (SCB) and the Swedish National Travel Survey. Being aware of the epistemological debate concerning the contradictions in combining the binary gender categories approach, and gender as a 'container', in relation to a more fluid understanding of gendered power relations and the 'doing' of gender (West and Zimmerman 1987), I will focus in this chapter on the binary sex categories of 'women' and 'men' in order to illustrate the importance of what statistical patterns of behaviour and the preferences of women and men can contribute regarding the analysis of everyday travel choices and climate concerns. I will, however, use the concept of gender. The text explores the extent to which gender differences in transport behaviour matter for climate sustainability for transport by providing statistics from the National Travel Survey and the most recent Swedish National Attitude Survey, showing gender differences regarding measures for the sustainability transition of the transport sector. The text further presents the results of an analysis of whether gender is included or mentioned in national policy decision documents, together with results from a previous survey on gender representation in decision-making bodies for transport.

My overall conclusion from what I present is used to argue for the importance of gendering transport policy and planning. The results of my analysis demonstrate the importance of moving forward on gender equality targets for the transport sector, not only for the sake of gender equality in the transport sector itself but also as part of gender equality for sustainable development.

TRANSPORT POLICY-MAKING

My professional work and this text are both rooted in the policy targets for the transport sector on gender equality and climate change. In the year 2000, an explicitly stated gender-specific focus was decided upon in the overall target of Swedish transport policy. This overall target is phrased as to "ensure an economically efficient and long-term sustainable transport for citizens and businesses throughout the country" (Prop. 1997/98:56; Prop. 2008a/09:93). The target is divided into two main parts, in which the first—the functional target—seeks to create accessibility with the intention that design, function and use of the transport system should help to give everyone basic accessibility to good-quality transportation that is usable in everyday life. This functional target states that *"the transport system should respond equally to women's and men's transport needs"*. The second part of the policy—the considerations targets—states that the targets of increased accessibility should be achieved together with road safety and improved environmental performance, as well as improved health. Regarding climate objectives, in 2008 the Swedish Parliament adopted a vision of zero net emissions of greenhouse gases to the atmosphere in Sweden by 2050 (Prop. 2008b/09:162). This was further strengthened by a new climate law which came into force on 1 January 2018, stipulating that gaseous emissions from domestic transport (except air) shall be 70% lower in 2030 compared with 2010 (Prop. 2016/17:146).

Pertinent and reliable statistics, data and measures are all essential components of policy-making and achieving any policy target. The necessary data is quite accessible in a country like Sweden, which provides open access to data on a plenitude of issues by Statistics Sweden—including on transport behaviour divided by gender. Even so, my experience is that very little gendered data is used in general transport policy. One might discuss what exact data to provide, what indicators to use and so on, but the basic level is to at least present gendered data to inform policy decisions. My experience while working as a professional planner is that data on travel behaviour disaggregated by sex is rarely used in general, nor is it present in any discussions on transport policy.

The European Institute for Gender Equality (EIGE) presents a Gender Equality Index providing a measure across member states of the complex concept of gender equality. It measures gender gaps within a range of areas relevant to the EU policy framework (work, money, knowledge, time, power, health, violence and intersecting inequalities). The Gender Equality Index assigns scores for member states, between 1 (total inequality) and

100 (full equality). The variation between member states is considerable, ranging from Greece's latest score of 50 to Sweden's score of 82.6, making Sweden an interesting case (EIGE 2017).

Sweden is even more interesting regarding transport policy issues since the country also has a (comparatively) high representation of women in transport decision-making. Among the 28 EU member states, Finland and Sweden have the highest representation of women among transport policy-makers and in transport administration (EIGE 2012). It is therefore particularly interesting that Swedish national transport policy, although recognizing the importance of the gender aspect by its specific gender target, shows few results (e.g. Swedish Transport Analysis 2013). A conclusion from this observation is that an equal gender representation in numbers does not necessarily guarantee gender-equal outcomes (Dymén et al. 2017).

The concept of gender mainstreaming has been embraced internationally as a policy strategy towards realizing gender equality. Gender mainstreaming has also been chosen by the Swedish parliament as the strategy for achieving gender equality (Prop. 1993/94:147). This method for achieving gender equality is not without criticism, for example, in how gender mainstreaming has transformed gender policy into administration and different forms of administrative techniques (Rönnblom 2011), and for being mismatched with the management control systems used by national transport authorities (Wittbom 2009). One possible explanation for the failure to make serious progress is that transport policy has not managed to develop tangible and coherent explanations for the problems to be solved (Rönnblom and Alnebratt 2016). That is to say, in this context, the connection between gender equality and problems relating to climate change and environmental sustainability is not well understood. This was also the conclusion of an evaluation of how the transport policy goal of gender equality was considered in the planning of transport infrastructure investments for the period 2010–2021 (Smidfelt Rosqvist et al. 2010). In this study (which will be further used in the text), officials from all levels (municipality, regional and national) responsible for infrastructure planning could not make sense of the gender equality target in relation to the transport planning process or their daily work.

However, since gender mainstreaming is the stipulated method for gendering the Swedish transport sector, I chose to use it and thus visualize and analyse the differences between women and men that affect transport policy-making with a special focus on sustainability. Thus, in the following, I present the data on the possible fit between gendering the transport sector and achieving its climate sustainability targets based on:

- gender differences in transport behaviour;
- gender differences in attitudes towards sustainability for transport;
- gender references (missing) in national policy decision documents;
- gender representation in decision-making for the transport sector.

SIGNIFICANT DIFFERENCES IN TRAVEL BEHAVIOUR

Taking a deepened understanding of gendered differences in travel behaviour seriously, the following section will consider some statistics that are rarely presented in full in relation to the making or support of transport policy. Data on travel patterns is derived from the Swedish National Travel Survey (RVU Sweden 2011–2014), which addresses people's daily travel, the dates and times when they travel, the modes of transport used and the purposes of their trips. The most recent study was conducted on a daily basis during the period 2011–2014; it encompassed the entire Swedish population in the age range of 6–84, and was conducted through telephone interviews. The survey was conducted over a period of four years and includes 39,280 interviews, corresponding to a response frequency of 42%. The database includes 57,577 car trips, 11,523 trips by public transport, 7074 by bike, 23,835 on foot and 1554 by other modes, adding up to a total of 101,563 trips.

Number of Trips

On average, 2.62 trips are made per person per day in Sweden (Table 4.1). Most are for leisure purposes (41%) followed by work/school (27%). Men make slightly fewer total trips but more trips for work/school, business and other purposes (Table 4.1). The greatest difference is for business, where men make almost twice the number of trips compared to women. Women make more trips for leisure (10%) and shopping/services (11%) than men.

More than half the number of daily trips (1.45, making up 55%) are made by car. Men make more trips by car (11%) and other modes (not including car, public transport, bike or walking) (27%), while women make more trips by public transport (23%), bike (4%) and walking (27%) (Fig. 4.1).

Transport Distances

Since travel patterns are often presented as percentages, women are often described as using more public transport and doing more shopping and service trips. In terms of distance travelled, however, there is no purpose/

Table 4.1 Average number of trips per person per day

	Work/school	Business	Shopping/service	Leisure	Other purposes	Total
Men	0.74	0.15	0.57	1.02	0.12	2.60
Women	0.69	0.08	0.63	1.12	0.12	2.63
Share trip purposes						
Men	29%	6%	22%	39%	5%	100%
Women	26%	3%	24%	42%	5%	100%
Men make more/less trips than women	7%	98%	−10%	−9%	4%	−1%
Women make more/less trips than men	−7%	−49%	11%	10%	−4%	1%

Source: RVU Sverige (2011–14)

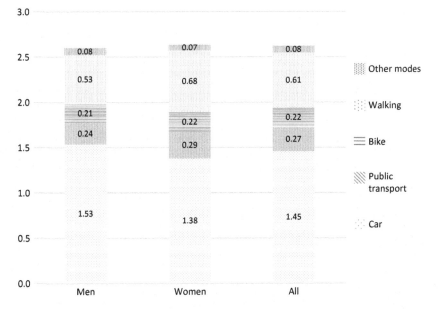

Fig. 4.1 Average number of trips per person and day split on mode of transport

activity where women outstrip men. Men travel further for all purposes, including those where women make more trips and, in some cases, have a higher share of the total than men do (Fig. 4.2).

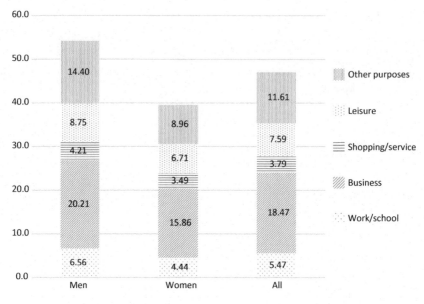

Fig. 4.2 Average total distance travelled (kilometre) per person per day split on trip purpose

Table 4.2 Average distance travelled per person per day (kilometre)

	Car	Public transport	Bike	Walking	Other modes	Total
Men	33.20	6.62	0.72	0.92	12.68	54.14
Women	23.17	6.59	0.50	1.06	8.15	39.47
Men make more/less than women	43%	0%	44%	−13%	56%	37%
Women make more/less than men	−30%	0%	−31%	15%	−36%	−27%

Source: RVU Sverige (2011–14)

When it comes to distance travelled, the car accounts for an even higher proportion of travel (60%) than reported for the number of trips. Men travel more kilometres (43%) by car than women, where using the car as a mode includes travelling as either driver or passenger. It is worth noting that there is no difference in kilometres travelled by public transport between women and men (Table 4.2). Men also travel more kilometres by bike (44%) and 'other' modes (56%). The only mode for which women travel more kilometres than men is for walking (15%).

Summing Up the Travel Differences

To summarize, men and women make approximately the same number of trips, but men travel longer distances than women do. This means that the total average distance travelled differs significantly between men and women. This is in line with the results of Carlsson-Kanyama et al. (1999) and Polk (2003), who analysed Swedish travel survey data from the mid-1990s. Little seems to have changed, which was also the conclusion in the assessment of Swedish national transport policy (Swedish Transport Analysis 2013), in which the results of two national travel surveys were compared: RES 2005–2006 and RVU Sweden 2010–2011. The assessment presented in the report concludes that the differences between men's and women's travel behaviour have remained constant over the years. Men still travel further than women and they *drive* twice the distance compared to women. One reason that has often been used to explain this gender difference in car use is differential access to the household car. This was refuted by Transport Analysis (Swedish Transport Analysis 2013). The gender-differentiated use of the car also persists in households with more than one car.

Similar patterns also occur in studies from countries such as Norway (Hjorthol 2008) and Germany (Scheiner et al. 2011). The most marked differences appear in work-related travel. Men on average make considerably more business trips—almost three times the distance compared to women—and commute considerably longer distances (RVU Sweden 2011–2014).

ATTITUDES TOWARDS CLIMATE CHANGE MEASURES ALSO DIFFER

Data on attitudes presented here is mainly based on the latest Swedish attitude survey conducted on behalf of the Swedish Environmental Protection Agency, in which public attitudes towards different solutions to achieving climate objectives were studied (Swedish Environmental Protection Agency 2015). The survey by the EPA was conducted online during the second week of March 2015 with a nationally representative sample in terms of geographical origin, gender and age. The questionnaire was answered by 1010 people. Questions in the study included topics such as the following: willingness to drive differently, for example, eco-driving; changing from car to public transport/bike/walking; buying different vehicles, for example, smaller or differently fuelled and so on.

Data on public attitudes towards different solutions to achieving climate objectives clearly shows that women are more in favour of changing transport behaviour and solutions supporting a sustainable transport sector (Swedish Environmental Protection Agency 2015). This finding has also been presented in earlier studies (e.g. Transek 2006; Polk 2003), showing that women place more emphasis on environmental and 'soft' issues for the transport sector than men do. It has been known for a long time (e.g. Lindén 1994) that women are more environmentally concerned and express more criticism of automobility than men do. The most recent national surveys on climate change (Swedish Environmental Protection Agency 2009, 2015) show differences in knowledge and attitudes towards climate change, results that also appear in international survey studies, indicating that women are consistently more engaged in the climate issue and consider it to be more important than men do (e.g. European Commission 2009; World Bank 2009). Women are also more in favour of *implementing* measures that could improve the situation and are more inclined to change their own behaviour (Polk 2003; Swedish Environmental Protection Agency 2007). In fact, *especially* questions related to mobility and transport behaviour show differences between men's and women's responses (Swedish Environmental Protection Agency 2007). For example, 80% of women were willing to consider driving less to reduce CO_2 emissions compared to 66% of men. Also, 75% of women, but only 53% of men, stated that they were willing to increase their use of public transport to reduce CO_2 emissions (Swedish Environmental Protection Agency 2007). The same pattern is revealed for ridesharing as well as driving at slower speeds to decrease the impact on climate change. The differences between men's and women's answers to questions regarding transport issues showed larger gaps than for other areas, such as temperature adjustments connected to clothes washing or the indoor climate.

GENDER NOT VISIBLE IN TRANSPORT POLICY OR DECISIONS

As the chair of *Nätverket för kvinnor i transportpolitiken* as a preparation for a speech with the title "Do transport planners care about gender equality?" at the Nordic transport conference Transportforum in 2013, I undertook two minor investigations of the most current major national committees to support national transport policy-making: the report "Commission on Fossil Free Road Transport 2013: 84" from the government bill *A Coherent Swedish climate and energy policy—Climate* (Prop.

2008b/09:162) and the "Investigation of capacity in the Swedish railway system" (Swedish Transport Administration 2012b: 005). The texts were simply searched, counting any references to or mention of gender (searching for the words woman/women, man/men, gender, gender equality). In neither of these documents were any policy-related gender references or analyses found.

Although not entirely omitting references to gender, the analyses by the Swedish Transport Administration (2012b: 005) and the Commission on Fossil Free Road Transport 2013: 84 seemingly rely on a very basic understanding of gender differences and present neither analysis nor discussions on the issue. Bearing in mind the specific gender target in the overall transport policy target, this must be regarded as very substandard.

The government bill *A Coherent Swedish climate and energy policy— Climate* (Prop. 2008b/09:162) outlines the strategic priority for Sweden, which is to have a vehicle fleet that is independent of fossil fuels by 2030, and for the vision for Sweden in 2050 to have a sustainable and resource-efficient energy system with no net emissions of greenhouse gases to the atmosphere. The appointed committee was commissioned to identify possible courses of action and measures to reduce the transport sector's emissions and dependence on fossil fuels in line with the vision for 2050. These measures may concern all the aspects that are important for achieving the long-term priority for 2030 and vision for 2050 and are interesting in relation to the gendering of transport as well as sustainable development efforts.

The result was presented in the document "Commission on Fossil Free Road Transport 2013: 84"—a more than 1000 pages' thick final report. Searching this report for the words: women, men, feminine, masculine, sex and gender equality gives a couple of hits. In Chapter 2, the overall transport policy target referring to the gender equality part is described. In Chapter 4, dealing with "Uncertainties and alternative projections of future situations", women are mentioned in relation to statistics on births (!) and car ownership. In Chapter 10, again the specific target on gender equality is mentioned as existing. In Chapter 15 on "Impact assessment", gender equality is dealt with in the following text:

From section 15.18

Men today use cars more and public transport less than women (Traffic Analysis 2012i). A move towards a more transport-efficient society is expected to equalize these differences. (Newman and Beatley 2009)

This citation expresses some awareness or idea that the decisions made for the transport sector are related to gender equality. However, in none of the chapters presenting or discussing specific measures or actions are gender or gender equality discussed or even mentioned. Even though the gender equality target exists and there is some awareness of the *existence* of relationships, this is not reflected or described within the identified choices and measures to reduce the transport sector's emissions and dependence on fossil fuels.

In the "Investigation of capacity in the Swedish railway system" (Swedish Transport Administration 2012b: 005), nothing at all was found referring to women, men, feminine, masculine, sex or gender equality.

Furthermore, on the same theme, an in-depth study of the work preparing the suggested objectives for the Swedish infrastructure plan from 2008 showed major trivialization of gender issues when the overall impact assessment documents were analysed (Smidfelt Rosqvist et al. 2010). The project initiated by the *Nätverket för kvinnor i transportpolitiken* evaluated how the transport policy sub-target of gender equality was considered in the work process of the action planning (which is the regular recurrent process for developing national and regional infrastructure plans), as well as how the resulting action plans were estimated to contribute to the achievement of the gender equality objectives of the transport policy. The project consisted of three parts, which analysed: (1) whether the work processes have been equally balanced regarding gender representation during the planning, (2) whether any gender issues have been discussed in the work process and (3) if the completed plans—the results of the work processes—contribute to gender equality.

Swedish infrastructure plans are obliged to follow the regulations on gender mainstreaming. More particularly, this is done by assessing the different investment objectives of the national transport policy in impact assessments. Smidfelt Rosqvist et al. (2010) examined the overall impact assessment documents for 284 investment objects with respect to how they were gender mainstreamed. Throughout, the assessments of how the gender equality target had been fulfilled were very brief indeed, and the question was frequently trivialized as many answers were simply copied and pasted from the instructions. Most formulations of the impact assessments were identical to the general instructions from the guidance on effects assessment handed out by the Transport Administration. The standard wording "*The measure is, however, not expected to have any significant impact on gender equality*" was given in the instructions and used in approximately 80% of the impact assessment documents examined, while there was no report on what the process and considerations to arrive at such a conclusion had been. Indeed, very few assessments were based on a report or analysis of a specific investment object's impact on gender equality. The repeated and routine use of formulations copied from the official guide reveals a lack of knowledge and understanding among those who produce these plans, of what the possible gender impacts of infrastructural investments are. It is also a sign that it is not considered to be an important issue, or an issue that needs to be considered within the infrastructure investment institutions' normative context.

Representation in Decision-Making Far from Gender Equal

The Swedish Transport Administration has 36% female officials and 64% men in total. This distribution has been roughly constant over the last three years at the management level, where 38% of the management consists of female officials and thus 62% are men. Although there has been a decrease in the number of women in leadership positions, this is not recognized in the annual report, which simply states that efforts are being made to increase the number of women in leadership positions (Swedish Transport Administration 2013).

In 2014 (approximately six months before an upcoming election), the *Nätverket för kvinnor i transportpolitiken* conducted a survey of gender representation in political committees dealing with traffic decisions at the municipal level. The average female representation in our survey was 27%. Only 27 of the 144 counted municipalities had 40–60% women, which corresponds to (almost) every fifth municipality. All the other 117 municipalities in the survey had fewer than 40% women. No municipality had a clear majority of women (i.e. over 60%). Eleven municipalities had no women at all, and 40 had fewer than 20% women on their committees deciding on transport issues.

During 2008–2010, long-term infrastructure plans covering national and regional investments in new roads, railways and navigational infrastructure for the period 2010–2021 were developed by the responsible national transport agencies and regional authorities in Sweden. A study (Smidfelt Rosqvist et al. 2010) initiated by *Nätverket för kvinnor i transportpolitiken* investigated whether a gender balance was achieved in the development of these long-term infrastructure plans for 2010–2021 by studying the composition of the various groups involved in the infrastructure planning process. The study concluded that the representation of men and women in the process differed regarding different kinds of tasks. In a little under half of the groups studied, those participating in both the national and regional working and steering groups were dominated by men (8 out of 17). In 8 out of the 17 studied types of groups, there was a gender-balanced distribution of men and women (40–60%) but this was mainly in the groups that can be assumed to have less influence over the outcome. That leaves one type of group where women were in the majority, and this was the working and advisory groups on environmental assessment. In the steering groups, where the most important decisions can be assumed to be taken, there was no gender balance.

Discussion: Letting the Differences Matter

Introducing a specific gender target into Swedish transport policy was partly due to known facts on different mobility patterns, for example, the work by Merritt Polk (1998). As well as data presented in this chapter, several other studies from Sweden (e.g. Kronsell et al. 2015; Carlsson-Kanyama et al. 1999; Polk 2003; Gil Solá 2013) have analysed mobility patterns and concluded that women's mobility is less car-dependent and leads to lower CO_2 emissions than men's. Even though these conclusions on important gender differences that ought to be recognized and discussed in relation to mobility and climate sustainability have been known for years, little or no attention has been paid to these perspectives on transport decision-making at a national level (Smidfelt Rosqvist et al. 2010), as exemplified by the investigations of transport policy documents presented here.

The differences in travel behaviour are also significant for the policy and practice on actions towards a sustainable transport sector. The data on travel behaviour presented above, and in more detail in Kronsell et al. (2015), suggests that, overall, women's travel patterns are in less need of adjustment in order to achieve the much-needed and urgent climate transition. This conclusion is further strengthened by women exhibiting a greater acceptance of actions needed on the path towards a more sustainable transport sector, and being more open to changing their behaviour than men. Kronsell et al. (2015) conclude that these behaviour and attitude characteristics expressed by women ought to be recognized and applied, for example, through contesting prevailing norms, standards and methods, in order to achieve the climate goals for the transport sector. Their calculations of CO_2 emissions from travel show that, if the travel behaviour of women was the norm, that is, if on average men exhibited the same travel behaviour as women, the reductions in distance travelled by car needed for sustainability would in fact already be in place. Further ongoing research on this topic shows that, considering population growth and the specific targets for Sweden, women's average travel patterns also need some adjustment, although still much less than men's (Smidfelt Rosqvist and Winslott Hiselius 2018).

In a situation where the transport sector is in need of a transition towards low-carbon fuels and sustainability, actions to more actively involve a gender perspective in transport policy-making thus go beyond increasing the democratic quality of policy-making by involving women as decision-makers. Such actions might also improve the democratic quality of policy-making for overall sustainability because including a gender perspective on measures and decisions for the transport sector would include making a stronger case for a more sustainable and safer transport sector. Moving towards equal representation, however, is also an absolute minimum as part of the democratic process. It is particularly important for a gender perspective as this means making sure that women's and men's conditions, preferences and values are equally included in the decision-making—all the way from defining the problem and the generation of alternatives through to the establishment of actual plans and investments. At the same time, a meta-analysis of results on female board representation has found this to be positively related to both a firm's financial performance and its board activities (Post and Byron 2015) as well as ongoing research showing a correlation between women's representation in decision-making on municipality transport boards and the sustainability performance of local transport (Dymén et al. 2017). In cases where there is an explicit policy target of gender equality, it might be argued that representation on its own would indeed improve target performance. This is a hypothesis that it would be very interesting to investigate without simplifying. There are intersectional perspectives on transport behaviour and attitudes that might play a role as well.

One policy implication of the findings is that more emphasis should be placed on the relationships between travel behaviour, climate and sustainability objectives and gendering at all levels of transportation policy-making, as also suggested by Kronsell et al. (2015). While officials have a hard time understanding what gender equality means for the transport sector and for what they do (Smidfelt Rosqvist et al. 2010; see also Levin et al. 2016), more information and knowledge about the relationships between travel behaviour, climate and gendering would be very likely to improve the understanding and 'making it matter'. Very little of this is visible, however, in the national debate on transport policy. Even though differences in mobility and attitudes are not contested as such, the debate often oversimplifies the differences that do exist. Differences in travel behaviour, for example, are often reduced to "women travel more by public transport and less by car", without reflecting upon either the magnitude of the differences or the definition of 'more' or 'less'. For example, in terms of the question at hand, it might be more accurate to talk about distance travelled than the number of trips, or the other way around. Depending on the reason for which the differences are presented, different indicators or measures should (and need to) be used and referred to.

It is obvious that the differences between men's and women's car usage at an overall level might suffice to form a challenge to climate change models. At the same time, one must acknowledge the tremendous complexity in the relationship between travel patterns, sustainability and gendering (Kronsell et al. 2015). This point is also discussed by Hanson (2010), who suggests research to synthesize thinking about gender and mobility, in both quantitative and qualitative approaches and across places. Even so, I would argue that the most urgent need for change is to disseminate knowledge (new and existing), to increase the level of competence among policy-makers and planners in the transport sector, contest prevailing norms and raise consciousness about gender impacts.

The transition to a gender-equal transport sector is a major task and, although not enough by itself, one early step on this road is to present the obvious overall gender differences regarding transport behaviour, attitudes and power distribution.

REFERENCES

Carlsson-Kanyama, A., Linden, A.-L., & Thelander, Å. (1999). Insights and applications gender differences in environmental impacts from patterns of transportation: A case study from Sweden. *Society & Natural Resources, 12*(4), 355–369.

Commission on Fossil Free Road Transport. (2013). Fossilfrihet på väg. SOU 2013: 84.

Dymén, C., Langlais, R., & Cars, G. (2013). Engendering climate change: The Swedish experience of a global citizens' consultation. *Journal of Environmental Policy & Planning, 16*(2), 161–181.

Dymén, C., Hiselius, L., Kronsell, A., & Smidfelt Rosqvist, L. (2017). Gender equal representation and sustainability in the transport sector. Paper presented at the conference gendering smart mobility 24–25 August 2017, Oslo.

EIGE, European Institute of Gender equality. (2012). *Review of the implementation in the EU of area K of the Beijing platform for action: Women and environment – Gender equality and climate change.* isbn:978-92-9218-024-9.

EIGE, European Institute of Gender Equality. (2017). *Gender equality index 2017 – Measuring gender equality in the European Union 2005–2015.* https://doi.org/10.2839/707843. isbn:978-92-9470-297-5.

European Commission. (2009). *European's attitude towards climate change.* Brussels: EU Commission.

European Commission. (2011). *Roadmap to a single European transport area: Towards a competitive and resource efficient transport system.* Brussels: EU Commission.

EEA, European Environment Agency. (2017). *Analysis of key trends and drivers in greenhouse gas emissions in the EU between 1990 and 2015* (EEA report no 8/2017). isbn:978-92-9213-861-5, issn:1977-8449. https://doi.org/10.2800/121780.

Gil Solá, A. (2013). *På väg mot jämställda arbetsresor?* (Towards gender equal commuting?). Doctoral thesis, Gothenburg University.

Hanson, S. (2010). Gender and mobility: New approaches for informing sustainability. *Gender, Place and Culture, 17*(1), 5–23.

Hjorthol, R. (2008). Daily mobility of men and women: A barometer of gender equality. In T. P. Uteng & T. Cresswell (Eds.), *Gendered mobilities* (pp. 193–210). Aldershot: Ashgate.

Kronsell, A., Smidfelt Rosqvist, L., & Winslott Hiselius, L. (2015). Achieving climate objectives in transport policy by including women and challenging gender norms: The Swedish case. *International Journal of Sustainable Transportation, 10*(8), 703–711.

Kurian, A. K. (2000). *Engendering the environment? Gender in the World Bank's environmental policies.* Aldershot: Ashgate.

Levin, L., Faith-Ell, C., Scholten, C., Aretun, Å., Halling, J., & Thoresson, K. (2016). *Att integrera jämställdhet i länstransportplanering* (Integrating gender equality into regional transport planning). K2 Research 2016: 1.

Levy, C. (2013). Travel choice reframed: 'Deep distribution' and gender in urban transport. *Environment & Urbanization, 25*(1), 47–63.

Lindén, A.-L. (1994). *Människa och miljö* (Human and environment). Stockholm: Carlssons.

May, A. D. (2013). Urban transport and sustainability: The key challenges. *International Journal of Sustainable Transportation, 7*(3), 170–185.

Moriarty, P., & Honnery, D. (2013). Greening passenger transport: A review. *Journal of Cleaner Production, 54,* 14–22.

Newman, P., & Beatley, T. (2009). *Resilient cities: Responding to peak oil and climate change.* Washington, DC: Island Press.

Nilsson, L. J., Khan, J., Andersson, F. N. G., Klintman, M., Hildingsson, R., Kronsell, A., Pettersson, F., Pålsson, H., & Smedby, N. (2013). *I ljuset av framtiden – styrning mot nollutsläpp 2050* (In the light of the future: Governing towards zero emissions 2050). LETS 2050-report, Lund University. Available online http://www.lth.se/fileadmin/lets2050/Rapporter_o_Abstracts/130831_Slutrapport_SE.pdf. Accessed 2 Sept 2015.

Polk, M. (1998). *Gendered mobility: A study of women's and men's relations to automobility in Sweden.* Göteborg: Institutionen för omvärldsstudier av människans villkor, avdelningen för humanekologi, GU, Doctoral thesis 1998.

Polk, M. (2003). Are women potentially more accommodating than men to a sustainable transport system in Sweden? *Transportation Research Part D, 8,* 75–95.

Post, C., & Byron, K. (2015). Women on boards and firm financial performance: A meta-analysis. *Academy of Management Journal, 58*(5), 1546–1571.

Prop. (1993/94:147). *Jämställdhetspolitiken: Delad makt – delat ansvar.* Swedish government bill on gender equality: Shared power – Shared responsibility.

Prop. (1997/98:56). *Transportpolitik för en hållbar utveckling.* Swedish government bill on transport policy for a sustainable development.

Prop. (2008a/09:93). *Mål för framtidens resor och transporter.* Swedish government bill on targets for future travelling and transport.

Prop. (2008b/09:162). *En sammanhållen klimat- och energipolitik – Klimat.* Swedish government bill on a coherent climate and energy policy – Climate.

Prop. (2016/17:146). *Ett klimatpolitiskt ramverk för Sverige.* Swedish government bill on a climate policy framework for Sweden.

Rönnblom, M. (2011). What's the problem? Constructions of gender equality in Swedish politics. *Tidskrift för genusvetenskap, 2–3,* 33–56.

Rönnblom, M., & Alnebratt, K. (2016). *Feminism som byråkrati: jämställdhetsintegrering som strategi* (Feminism as bureaucracy: Gender mainstreaming as strategy). Stockholm: Leopard förlag.

RVU Sweden. (2011–2014). *The national travel survey, RVU Sverige* (Sweden) 2011–2014. Swedish Transport Analysis and SCB.

Scheiner, J., Sicks, K., & Holz-Rau, C. (2011). Gendered activity spaces: Trends over three decades in Germany. *Erkunde, 65*(4), 371–387.

Smidfelt Rosqvist, L. (2009). Post-script on traffic systems for an improved city environment. *Journal of World Transport Policy & Practice, 15*(3), 54–55.

Smidfelt Rosqvist, L., & Wennberg, H. (2012). Harmonizing the planning process with the national visions and plans on sustainable transport: The case of Sweden. *Procedia – Social and Behavioral Sciences, 48,* 2374–2384.

Smidfelt Rosqvist, L., & Winslott Hiselius, L. (2018). Understanding high car use in relation to policy measures based on Swedish data. *Case Studies on Transport Policy.* https://doi.org/10.1016/j.cstp.2018.11.004.

Smidfelt Rosqvist, L., Dickinson, J., Billsjö, R., Nilsson, A., & Söderström, L. (2010). *Jämställdhet i infrastrukturplaneringen – en utvärdering* (Gender equality in infrastructure planning: An evaluation). Trivector Report 2010:38.

Swedish Environmental Protection Agency. (2007). *Genusperspektiv på allmänhetens kunskaper och attityder till klimatförändringen* (tidigare växthuseffekten) (Gender perspective on public knowledge and attitudes towards climate change (formerly greenhouse effect)). ARS P0924.

Swedish Environmental Protection Agency. (2009). *Allmänheten och klimatförändringen* (The public and climate change). Report No. 6311.

Swedish Environmental Protection Agency. (2015). *Allmänheten och klimatförändringen 2015* (The public and climate change 2015). Rapport 2015-05-22.

Swedish Transport Administration. (2012a). *Delrapport Transporter. Underlag till färdplan 2050* (Interim report for transport: Documentation for Road Map 2050). Trafikverket publication 2012:224, Borlänge.

Swedish Transport Administration. (2012b). *Investigation of capacity in the Swedish railway system: Suggested solutions for the years 2012–2021*. Trafikverket publication 2012: 005.

Swedish Transport Administration. (2013). Annual Report: 2013 https:// trafikverket.ineko.se/Files/sv-SE/11769/RelatedFiles/2014_054_the_swedish_transport_administration_annual_report_2013.pdf

Swedish Transport Analysis. (2013). *Uppföljning av de transportpolitiska målen* (Follow-up of the transport policy objectives). Report No. 4.

Transek. (2006). *Mäns och kvinnors resande. Vilka mönster kan ses i mäns och kvinnors resande och vad beror dessa på?* (Men's and women's travelling: What patterns can be seen in men's and women's travelling and what are they due to?). Report No. 51.

West, C., & Zimmerman, D. H. (1987). Doing gender. *Gender & Society, 1*(2), 125–151.

Wittbom, E. (2009). *Att spränga normer – om målstyrningsprocesser för jämställdhetsintegrering* (Breaking norms: On management by objectives for gender mainstreaming). Stockholms universitet, Samhällsvetenskapliga fakulteten, Företagsekonomiska institutionen.

World Bank. (2009). Public attitudes toward climate change: Findings from a multi-country poll in *World development report 2010: Development and climate change*. Washington, DC: World Bank.

CHAPTER 5

How to Apply Gender Equality Goals in Transport and Infrastructure Planning

Lena Levin and Charlotta Faith-Ell

Introduction

Planners and politicians working on gender equality in transport and infra-structure need to adapt their work to the plan or project currently at hand. In this case, this entails choosing relevant goals (from the 2030 Agenda for Sustainable Development) and adapting them to their everyday work; for example, the goal of enhancing the use of enabling technology, particularly information and communications technology, to promote the empowerment of women, ensure access to paid employment and eliminate significant gaps between men and women in the labour market (UN 2015a, b, Goal 5). For politicians, this also entails adopting and strengthening sound policies and enforceable legislation promoting gender equality and the empowerment of

L. Levin (✉)
VTI, Linköping, Sweden
e-mail: lena.levin@vti.se

C. Faith-Ell
WSP, Stockholm, Sweden

Estonian Environment Institute, Tartu, Estonia
e-mail: charlotta.faith-ell@wsp.com

© The Author(s) 2019   89
C. L. Scholten, T. Joelsson (eds.), *Integrating Gender into Transport Planning*, https://doi.org/10.1007/978-3-030-05042-9_5

all women and girls at all levels. This chapter presents a model in which gender equality goals are implemented in transport planning. We propose a structure and systematics based on previous research into gender equality and provide suggestions on how to systematically integrate gender into transport planning. We argue that both organizational and political works are needed (cf. Faith-Ell and Levin 2013; Levin and Faith-Ell 2014).

According to Chantal Mouffe (2005, 2013), politics can be seen as a process of dynamic identification and negotiation between social forces engaged in a struggle for power. Mouffe believes that the relevant categories can never be pinned down—there are countless ways to look at concepts of identity and identification and, during historical power processes, certain conventions have gained precedence over other ways of viewing the world—and it follows that no identity can be anchored to an original essence or core (Mouffe 2005, 2013). Large-scale projects have often resulted in failure because they have tried to solve problems from the top down when they also need to take account of power struggles and lack of power: for example, the need to increase people's agency, whether concerning matters of employment, market production, or household reproduction (Sharp et al. 2003).

A common criticism of gender mainstreaming is its lack of standardization of goals, procedures and methods, as well as its frequent lack of unambiguous self-definition (cf. Eveline et al. 2009). The present contribution concerns efforts to bridge the gap between research and planning. Over the past decade, we have conducted research close to practice; for example, in the road-building, railway planning and public transport planning areas. This means that we have received instant feedback from practitioners and tried to adapt our model to fit their needs, and explains why we have been influenced by the impact assessment (IA) procedure in developing the model. Furthermore, we find that our model has attracted attention from policy-makers outside our research projects, and that our findings are now sought out by county councils and regional transport planners. Our previous research has focused on Swedish gender objectives. In light of the development of the gender goals in the UN's 2030 Agenda for Sustainable Development (henceforth, '2030 Agenda'), this chapter concentrates on how Goal 5 of the 2030 Agenda can be applied in transport planning and design. In our theoretical work, we see the transport system as facilitating the achievement of gender objectives, since the design of the transport system is central to upholding various parts of the daily life of citizens, both women and men, that is, accessibility in a broad sense.

In this chapter, we investigate how policy goals concerning gender can be operationalized in transport planning and thus contribute to gender mainstreaming in the transport sector (cf. Christensen et al. 2007; Woodward 2003). One part of the background is our previous research in Sweden. We have also based the chapter on the 2030 Agenda (UN 2015b), and we suggest how to implement the Agenda in gender mainstreaming work in plans and projects.

The chapter is structured as follows. First, we present the background to our research into gender and transport planning. This is followed by an overview of gender in transport politics, the goals of gender mainstreaming and the connections to an IA procedure.

We then present our analysis, with suggestions on how to operationalize goals and objectives and adapt to targets and indicators in the field of transportation planning. In this section, we suggest and discuss actions to promote gender mainstreaming in transportation planning and methods that can be adapted to this area. The analysis is linked to theories of justice and how the politics of transport planning can be made more equal. When we use the word 'political', we mean that any difficulties encountered refer to divergent interests and to struggles over questions of justice and redistribution between groups. Here we are inspired by the political theorist Chantal Mouffe (2005, 2013) and her distinction between *politics* and the *political*: politics concerns institutions, such as parties, parliaments and public administrations, while the political concerns battles, conflict, negotiation and distribution issues. The chapter concludes by discussing the implications of the present research into practice and theory.

BACKGROUND

Sweden has a long tradition of gender equality work, and the Swedish parliament and government have introduced national gender equality goals into transport politics (Smidfelt Rosqvist, this volume). However, evaluations have identified difficulties in implementing these goals in practice; for example, a significant gender pay gap, lack of women in top management positions in businesses, and a larger share of the unpaid labour of caring for children and relatives, which disfavour women's lifetime earnings from paid labour (Alnebratt and Rönnblom 2016; Göransson 2007). We believe that these inequalities are due to deep-rooted politics in all fields and political assemblies, and more practice-oriented planning and construction teams.

Annual evaluations of Swedish transport policy objectives[1] show that differences between men's and women's travel have remained constant over the past 20 years. Men spend more time on travel linked to business or education and less time on travel for household purchasing and service matters. Compared with women, men make approximately 20% fewer trips linked to unpaid domestic care and work. Men drive cars almost 80% further than women do, according to evaluation of the gender equality objective in Swedish transport policy (Trafikanalys 2017: 28–29; cf. Smidfelt Rosqvist, this volume). Also, evaluations conducted by the Swedish National Council for Crime Prevention (BRÅ)[2] show that the number of people who feel insecure in their residential areas in the evening has increased over the past few years. There are great differences between women and men in this regard: over 30% of women report feeling so much insecurity that they avoid going out, versus 9% of men experiencing the same kind of insecurity. Experiencing such anxiety can mean that people choose not to travel by public transport, walking or cycling (Trafikanalys 2017: 17).

Because there are differences between men's and women's travel patterns, evaluations state (Trafikanalys 2017: 30–31) that it is important to capture both men's and women's experiences of the decision-making processes that steer transport system development. Among other things, it is important that the transport system's decision-making assemblies have equal representation of women and men. There are at least two reasons for this: one is to serve as a model by demonstrating that it is possible for both men and women to have careers in the transport sector, while the second is to help the transport sector become gender equal. In recent years, the proportion of women in management and government agencies in the transport sector has increased to 40–50%. However, equal representation is not enough to contribute to gender mainstreaming. It is necessary to ensure that both women's and men's experiences are considered in all transport sector decision-making. The design of the transport system depends largely on decisions made at the regional and local levels. There is still a lower representation of women at this level (on average 30–35% in 2016), and in some municipalities there are no women in the decision-making bodies responsible for the transport sector (Dymén et al. 2017). The transportation sector is interconnected with the political arena and the decision-making processes occurring there, meaning that the political agenda shapes transport planning and is an important component of implementing gender policy. Gender policy should permeate all levels of community planning. However, different goals have different statuses in politics. In some fields, such as the transportation sector, gender equality and other social issues are more under-emphasized than otherwise, as demonstrated by, for example, Wanna Svedberg (2014) in her legal study of Swedish transport policies.

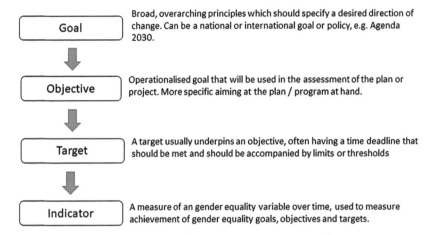

Broad, overarching principles which should specify a desired direction of change. Can be a national or international goal or policy, e.g. Agenda 2030.

Operationalised goal that will be used in the assessment of the plan or project. More specific aiming at the plan / program at hand.

A target usually underpins an objective, often having a time deadline that should be met and should be accompanied by limits or thresholds

A measure of an gender equality variable over time, used to measure achievement of gender equality goals, objectives and targets.

Fig. 5.1   The goals–objectives–targets–indicators hierarchy used in our model

In our research on the Swedish transport planning context, we have suggested that the transport system can facilitate gender mainstreaming in society, since it can provide mobility that allows citizens to access arenas such as education and paid work. It can also broaden the opportunities for more adults, both men and women, to share housework. We believe that one point of departure in integrating and assessing gender equality in transport is gender equality goals, and we accordingly highlight the process of adopting these goals into planning practice. Basing this process on gender equality goals will bolster its legitimacy because it can be traced (tiered) to national or global objectives, that is, the operationalized goals (see Fig. 5.1 above). Furthermore, this approach will contribute to the achievement of these goals.

Overview of Gender in Transport Politics

To understand the conditions and critical aspects of gender mainstreaming, we should bear in mind that gender refers to power relations in, for example, organizations, social practices and institutions and that differences are often assumed to exist between women and men (Connell 2006; Fainstein and Servon 2005). To be precise, institutions historically dominated by men (e.g., in the transport sector) reflect masculine norms and values, meaning that the male (travel) agenda remains the norm (cf. Balkmar 2012; Joelsson 2013; Kronsell 2005; Mellström 2002). However, as more women enter transportation politics and the planning profession, it is no longer merely an issue of the numerical dominance of men; rather, it is the dominance of male norms in the theories and principles learned in education and practised in everyday work (Forsberg and Lindgren 2015; Hirdman 2003; Sandercock and Forsyth 2005).

Here gender is considered to be a process or routine enacted through everyday actions and interactions, and not as the pre-defined expression of underlying sex identity grounded in biology (West and Zimmerman 1987). As humans, we are born into a world already linguistically and conceptually organized. By participating in various activities and contexts in which certain categories and concepts are used, we learn to view the environment, people, things and actions in such a way that they become meaningful and manageable. New categorizations are also generated in interaction with others, especially if this interaction is necessary to establish and maintain relationships and coordinate activities with other people. Categorization entails certain rights, responsibilities and expectations, serving as a social mechanism or structuring principle to maintain the social order (Goffman 1971, 2004; Jayyusi 1984). 'Doing gender' (West and Zimmerman 1987) means that individuals create similarities and differences between each other that are not natural, essential or biological. Talking about girls and boys and women and men does not mean that there is only one way to interpret these categories, even though societies, cultures and humans may jointly create and sometimes agree on common aspects applicable to a category. Many of the most common categorizations of people (e.g., men and women, young and old, and natives and immigrants) become clear to individuals through meeting with authorities or institutions. Candace West and Don Zimmerman's (1987) theoretical shift towards 'doing' instead of 'being' conceptualizes gender as an accomplishment rather than a fixed attribute that is politically useful. Gender equality (Swedish, *jämställdhet*) is therefore not a stable concept but must be continuously (re)processed. Having this in mind improves our understanding of how change affects different levels of government and social institutions as well as everyday life.

Studies demonstrate that historical structures may persist and hold back gender equality work at the regional level (Forsberg and Lindgren 2015). In their research, Gunnel Forsberg and Gerd Lindgren (2015) found that the leaders of a new development programme for sustainable growth expanded their power base through male networks. These historical networks (e.g., lodges) as a rule require the appointment of members and they largely exclude women. The influence of these male networks on regional development policy seems to persist even today, meaning that new activities and clusters are difficult to develop. There are women in politics and in positions that require representative and appointment procedures at a national level (cf. Göransson 2007), but the number of women decreases when peer-selection methods (Tienari et al. 2013) are used, as in the private sector and regional clusters (Forsberg and Lindgren 2015).

Gender equality work itself incites tensions in the political arena. The story of gender equality and gender studies in Sweden is in many respects a success story, although it is not unproblematic. It is a story in which concepts such as feminism, gender and equality have been negotiated (cf. Alnebratt 2009). We can note here that the female/male binary has usually been the basis of political projects.

Ana Gil Solá (2016) has investigated gender relationships in terms of gender contracts in households (cf. Hirdman 1993, 2003). Based on 20 in-depth interviews with parents of small children, Gil Solá identified three distinct types of gender contracts: the traditional, gender-equal, and mixed-gender contracts. "Depending on the type of contract, households will handle the constraints imposed by surrounding socio–spatial structures and exploit opportunities in different ways, leading to different activity spaces on the labour market. When defining the gender contract of a household, car use plays a particularly important role" (Gil Solá 2016: 39). Households with traditional gender contracts prioritized men's activities, while households with gender-equal contracts prioritized equivalent activities for women and men, resulting in more flexible arrangements based on the household members' negotiations. The mixed-contract households ended up with some cases of unequal and some cases of equal opportunities on the labour market.

Apparently, there is tension between the legal definitions of sex and gender equality that politicians work with, and the more open identity categories that form and transform people's everyday lives.

The binary division of gender can be seen as two complementary halves, male and female, that are equal in a relationship; however, the two are often, instead, characterized as opposites, with the male as the norm and the female in the othered position. In sociology and anthropology, the study of gender considers not simply the division and problematization of the categories of male and female, but rather the whole set of meanings that the sexes enact in societies or particular situations. The operationalization of gender in practice takes up these meanings, organizes them as masculinity and femininity, and matches them with male or female bodies. This can be connected to political standpoints on travel behaviour, based on studies identifying differences in men's and women's travel trajectories and then drawing conclusions that men and women have different travel needs. Robert W. Connell (1996) used the term 'gender regimes' to refer to the gendered social practices exemplified by various institutions (e.g., schools and workplaces) and practices of everyday life (e.g., caregiving, teaching and travelling). However, the political process of equality needs to take account of various power structures dealing with sexuality (including gay/lesbian, bi and trans issues), ethnicity and other identity issues. In gender research, central issues are difficult to find in the political contexts in which gender equality is discussed. It seems particularly clear in transport policy that only one social category can be discussed at a time.

According to Emanuela Lombardo et al. (2010), fixing on a certain meaning is a common process in defining what is to be understood by gender equality: "When particular definitions of gender equality are created in a given context (for instance, equal opportunities or positive actions, or gender equality in politics) they can be fixed for some time. /.../ In this respect, the fixing of the concept of gender equality is the result of a discursive struggle" (Lombardo et al. 2010: 108). Definition is also an achievement in the gender struggle, meaning that gender equality has been enshrined in legal and political documents and has (in the best case scenario) been recognized and is no longer a contested goal.

In this study, we use gender equality as a concept because of attempts to discuss issues related to this concept and to propose action. There is an awareness of norm criticism but still a need to focus on gender equality in planning practice. The definition of gender equality as equality between binary sexes (i.e., male and female) can constitute the basis on which actors challenge other meanings and initiate processes of contestation that may result in broader interpretations of gender equality that are more inclusive or better account for the complexity of gender inequality (cf. Lombardo and Meier 2008). Joan Eveline et al. (2009) have pointed out that a challenge for gender mainstreaming is to find a strategy for addressing intersectional oppressions, such as when oppressive practices (e.g., sexist *and* racist) intersect in women's lives. They suggest that gender mainstreaming needs to incorporate the process of "translating declared goals into action" (Eveline et al. 2009: 199). This is the perspective we introduced in our previous research into Swedish transport planning (Faith-Ell and Levin 2013).

CONCEPTUALIZING GOALS OF GENDER MAINSTREAMING

Early on in our research into gender mainstreaming in transport planning, it became clear to us that much transport planning was based on clear objectives and goals. Although Sweden had an overarching transport policy goal of achieving "a gender-equal transport system", there were no clear objectives or goals for gender mainstreaming in transportation plans or projects (Smidfelt Rosqvist, this volume). However, at the same time, the country had national gender equality goals that could be used in transport planning.

Based on our research, we argue that gender equality goals can be used as a basis for gender mainstreaming in transport planning. We see policy goals as key components in constructing tools for building a transportation system that is available to and usable by all citizens. Our approach has been to operationalize existing goals and policies into objectives that can be applied as a basis for assessment or targets in planning. This approach was chosen because planners and politicians working on gender equality in transport and infrastructure can choose relevant goals and adapt them to their everyday work. Another reason for choosing this approach is that several international and national policies and agreements have set goals for gender mainstreaming and women's empowerment that can be used in transport planning. Although most of the available goals, objectives and policies are not legally binding and should essentially be viewed as expressions of political will, we suggest that they be used in gender mainstreaming regardless of their country of origin. The following are examples of goals and policies that could be used.

The Council of European Municipalities and Regions (CEMR) charter for equality of women and men in local life (henceforth, "CEMR charter"), launched on 12 May 2006 by the CEMR, is intended to encourage local and regional governments in Europe to commit to concrete measures for gender equality at all levels of local life (CEMR 2006).

The new 2030 Agenda for Sustainable Development and its 17 sustainable development goals (SDGs) officially came into force on 1 January 2016 (UN 2015b). Although gender equality and women's empowerment are formulated as a stand-alone goal (Goal 5), several of the other 16 SDGs also target gender equality.

The European Union (EU) has set up a framework for the Commission's future efforts to improve gender equality, called Strategic Engagement for Gender Equality 2016–2019 (European Commission 2015). The Gender Equality Index 2017 shows where Europe stands today. Gender equality is moving forward but progress is very slow. The EU's score is just four points higher than it was ten years ago, now 66.2 out of 100. Sweden's score is 82.6 while Greece is at the bottom with 50 points (EIGE 2017). Furthermore, several countries have national gender equality goals and policies that can be used in transport planning. Sweden's overall policy goal of gender equality is that women and men are to have the same power to shape society and their own lives. Based on this overall goal, six sub-goals have been adopted by the Swedish parliament and government (Government Bill 2005/06:155; Government Offices of Sweden 2016).

An overview of existing gender equality goals and policies identifies a single overarching goal, that is, "ending all forms of discrimination against all women and girls everywhere", which is common to all specific goals and policies. Table 5.1 summarizes how this is formulated in various sets of goals.

Based on our research, we have compiled five categories of gender equality goals that can be used in transport planning (Table 5.2). These categories have been derived from international and national policies and agreements establishing goals for gender mainstreaming and women's empowerment (Table 5.3). As evident from Table 5.3, which visualizes

Table 5.1 Summary of the overarching goal "ending all forms of discrimination against all women and girls everywhere"

2030 Agenda: 5.1. End all forms of discrimination against all women and girls everywhere
2030 Agenda: 5.c. Adopt and strengthen sound policies and enforceable legislation for the promotion of gender equality and the empowerment of all women and girls at all levels
CEMR charter: 1. Equality of women and men constitutes a fundamental right
CEMR charter: 2. To ensure the equality of women and men, multiple discrimination and disadvantage must be addressed
CEMR charter: 4. The elimination of gender stereotypes is fundamental to achieving equality of women and men
CEMR charter: 5. Integrating the gender perspective into all activities of local and regional government is necessary to advance equality of women and men
CEMR charter: 6. Properly resourced action plans and programmes are necessary tools to advance equality of women and men
EU: 5. Promoting gender equality and women's rights across the world
Sweden: The overarching goal of the gender equality policy is that women and men are to have the same power to shape society and their own lives

Table 5.2 Categories of gender equality goals in transport planning

A. Eliminate all forms of violence against all women and girls
B. Democracy and influence on decision-making
C. Equal opportunities for good health and personal development, including access to culture and leisure
D. Equal opportunities and conditions for education and paid work
E. Equal distribution of unpaid housework and provision of care

Table 5.3 Categorization of international and national gender goals and policies

CEMR charter				
UN 2030 Agenda for Sustainable Development				
EU Strategic Engagement for Gender Equality 2016–2019				
Sweden Gender Equality Goals				
A. Democracy and influence on decision-making	B. Eliminate all forms of violence against all women and girls	C. Equal opportunities for good health and personal development, including access to culture and leisure	D. Equal opportunities and conditions for education and paid work	E. Equal distribution of unpaid housework and provision of care
CEMR charter: 3. The balanced participation of women and men in decision-making is a pre-requisite of a democratic society	2030 Agenda: 5.2. Eliminate all forms of violence against all women and girls in the public and private spheres, including trafficking and sexual and other types of exploitation	2030 Agenda: 4.1. By 2030, ensure that all girls and boys complete free, equitable and quality primary and secondary education leading to relevant and effective learning outcomes	2030 Agenda: 5.5. Ensure women's full and effective participation and equal opportunities for leadership at all levels of decision-making in political, economic and public life	2030 Agenda: 5.4. Recognise and value unpaid care and domestic work through the provision of public services, infrastructure and social protection policies and the promotion of shared responsibility within the household and the family as nationally appropriate
2030 Agenda: 5.5. Ensure women's full and effective participation and equal opportunities for leadership at all levels of decision-making in political, economic, and public life	EU: 4. Combating gender-based violence and protecting and supporting victims	2030 Agenda: 4.2. By 2030, ensure that all girls and boys have access to quality early childhood development, care and pre-primary education so that they are ready for primary education	2030 Agenda: 5.4. Recognise and value unpaid care and domestic work through the provision of public services, infrastructure and social protection policies and the promotion of shared responsibility within the household and the family as nationally appropriate	2030 Agenda: 5.a. Undertake reforms to give women equal rights to economic resources, as well as access to ownership and control over land and other forms of property, financial services, inheritance and natural resources, in accordance with national laws
2030 Agenda: 5.a. Undertake reforms to give women equal rights to economic resources, as well as access to ownership and control over land and other forms of property, financial services, inheritance and natural resources, in accordance with national laws	Sweden: Men's violence against women must stop. Women and men, girls and boys must have the same right and access to physical integrity	2030 Agenda: 4.3. By 2030, ensure equal access for all women and men to affordable and quality technical, vocational and tertiary education, including university	2030 Agenda: 4.3. By 2030, ensure equal access for all women and men to affordable and quality technical, vocational and tertiary education, including university	Sweden: Equal distribution of unpaid housework and provision of care. Women and men must have the same responsibility for housework and have the opportunity to give and receive care on equal terms
2030 Agenda: 5.b. Enhance the use of enabling technology, in particular information and communications technology, to promote the empowerment of women		2030 Agenda: 4.5. By 2030, eliminate gender disparities in education and ensure equal access to all levels of education and vocational training for the vulnerable, including persons with disabilities, indigenous peoples and children in vulnerable situations	2030 Agenda: 4.5. By 2030, eliminate gender disparities in education and ensure equal access to all levels of education and vocational training for the vulnerable, including persons with disabilities, indigenous peoples and children in vulnerable situations	

(*continued*)

Table 5.3   (continued)

2030 Agenda: 5.c. Adopt and strengthen sound policies and enforceable legislation for the promotion of gender equality and the empowerment of all women and girls at all levels		Sweden: Equal health. Women and men, girls and boys should have the same prerequisites for good health and be offered medical assistance and care on equal terms	2030 Agenda: 8.5. By 2030, achieve full and productive employment and decent work for all women and men, including for young people and persons with disabilities, and equal pay for work of equal value	
EU: 3. Promoting equality between women and men in decision-making			2030 Agenda: 5.a. Undertake reforms to give women equal rights to economic resources, as well as access to ownership and control over land and other forms of property, financial services, inheritance and natural resources, in accordance with national laws	
Sweden: Equal division of power and influence. Women and men are to have the same rights and opportunities to be active citizens and to shape the conditions for decision-making			EU: 1. Increasing female labour market participation and equal economic independence	
			EU: 2. Reducing the gender pay, earnings and pension gaps and thus fighting poverty among women	
			Sweden: Economic equality. Women and men must have the same opportunities and conditions as regards education and paid work which give economic independence throughout life	
			Sweden: Equal education. Women and men, girls and boys shall have the same opportunities and terms when it comes to education, study options and personal development	

'the political' in our research, the goals predominantly relate to institutional conditions (e.g., politics and political arenas, labour market, education, health and care). In particular, the labour market emerges as a priority area, possibly due to the history of women's struggle for equality. Historically, this struggle has been about education rights for girls, women's labour, antenatal care and healthcare for women. In Sweden, gender equality has historically been handled by the Ministry of Labour, and political initiatives have concerned equality in working life (e.g., equal pay for women and men) and the establishment of gender research and opportunities for women to have academic careers (Alnebratt 2009; Alnebratt and Rönnblom 2016; cf. Hanson 2010).

An Objectives-Led Approach to Gender Impact Assessment

The process we have developed for integrating gender perspectives into transportation planning is called gender impact assessment (GIA), a process that is conceptualized based on the philosophy of impact assessment. We have taken inspiration from the fields of social impact assessment (SIA) and strategic environmental assessment (SEA), combining them into a tool or process for integrating gender equality into transport planning. The model can be described as an objectives-led *ex ante* evaluation of transport plans, projects and policies. The following is a brief description of the two main sources of influence on the outline of the model.

SEA is defined in many ways in the IA literature. One widely used definition is that "SEA is a systematic process for evaluating the environmental consequences of proposed policy, plan or programme initiatives in order to ensure they are fully included and appropriately addressed at the earliest appropriate stage of decision-making on a par with economic and social considerations" (Sadler and Verheem 1996: 27).

Two approaches can be identified in the SEA literature: objectives-led and baseline-led. Objectives-led SEA predicts whether or not a strategic action will help fulfil a range of SEA objectives. This means that the SEA objectives act as an independent 'sustainability/environmental yardstick' against which a strategic action can be tested (Therivel 2010). In contrast, baseline-led SEA starts from an existing environmental baseline and makes predictions about how the strategic action will change this baseline (Therivel 2010).

SIA is also defined in various ways. Our research has applied the definition of SIA articulated in the "International Principles for Social Impact Assessment", formulated by the SIA community as follows: "the processes of analysing, monitoring and managing the intended and unintended social consequences, both positive and negative, of planned interventions (policies, programs, plans, projects) and any social change processes invoked by those interventions" (Vanclay 2003: 5).

In SIA, the baseline-led approach to the evaluation of plans and projects is dominant because SIA was initially developed to focus on individual projects for the purpose of practice. This means that an objectives-led approach to assessment and evaluation has rarely been applied. However, based on our experience of strategic transport planning, we believe that gender equality goals and policies can be used in assessing strategic actions in the same manner as environmental goals. We argue that an objectives-led approach to GIA could further develop transport planning with regard to gender mainstreaming. Firstly, the application of an objectives-led approach to GIA could *improve the outcomes of transport planning*, the reason being that GIA, when linked to the strategic level of politics, provides a systematic tool for integrating gender equality goals into transport planning. Secondly, an objectives-led approach to GIA would *provide an idea* of whether or not various strategic actions are moving in the desired direction. This means that transport planners would be able to evaluate the *fulfilment of the 2030 Agenda goals* or national gender equality goals. Thirdly, this approach to GIA would *establish assessment criteria to be used for testing the effects on gender equality* of strategic actions of transport plans. Fourth and finally, GIA based on objectives would also *highlight potential goal conflicts.*

The four main functions of our GIA model are as follows (cf. Halling et al. 2016):

- to integrate gender equality into planning as a relevant consideration, that is, work towards gender mainstreaming;
- to assess the impacts of a plan or project, that is, inform decision-makers about the impacts of strategic choices in planning on various groups;
- to mitigate potential impacts, that is, find ways to ameliorate adverse negative impacts on affected groups; and
- to give guidance for subsequent stages.

As in all other IA tools, the process in the GIA model is based on well-defined steps (cf. Halling et al. 2016; IAIA 2016). The emphasis of GIA is on the process rather than the final GIA report. The GIA report nevertheless plays an important role, being one of several documents that serve as a basis for the plan and project decision-making. However, if the process is not carried out properly, the decision basis will be incomplete. The main steps of the IA process are

1. inception;
2. scoping of significant aspects and goals as well as identification of a basis for assessment, that is, GIA objectives;
3. baseline studies;
4. identification and involvement of relevant groups;
5. identification, prediction and evaluation of impacts of significant aspects based on the GIA objectives;
6. suggestion of measures for mitigating adverse impacts of the proposed action;
7. assessment of the impacts of the final design;
8. compilation and writing of the GIA report;
9. formulation of suggestions on the integration of gender measures into the subsequent planning; and
10. follow-up of identified significant impacts.

Operationalizing the Goals of Gender Mainstreaming

The goals of gender equality are often considered vague by transport planners and decision-makers since they are usually formulated in a general way. Furthermore, the goals are rarely adapted to the practice at hand. To be applicable in transport planning, gender equality goals need to be operationalized. By operationalization, we mean that the overarching theoretical goals are translated into more tangible objectives and indicators that fit the specific planning situation (Fig. 5.1). We have chosen to operationalize goals from the five-goal categories in Table 5.2 as objectives, targets and indicators in our work. Furthermore, it is important to note that one goal can be operationalized as several objectives if needed. The goals in Table 5.2 can be operationalized in many ways. The following are examples of how the five categories of gender equality goals (Table 5.2) can be operationalized as objectives and indicators.

Democracy and Influence on Decision-Making

All citizens should have the opportunity to influence public policy and decision-making by participating in democratic processes, that is, voting and in other ways having their voices heard in the procedures that create people's everyday living conditions. This also includes decisions about, for example, residential space and transportation systems. In transportation planning, this means that both men and women should be involved in all phases of transport planning and development (e.g., equal participation in project development, public consultation, and review).

A possible operationalization of this goal could be:

> There should be equal distribution of power and influence between women and men in decision and planning processes in infrastructure planning.

Based on our research, we argue that every new phase of transport planning should start with an analysis of the power and influence of different groups involved in the planning process. This analysis should cover the planning/project organization (i.e., proponents and consultants), competent authorities, stakeholders and the public group affected (e.g., transport system users). Box 5.1 describes examples of targets and indicators.

Eliminate All Forms of Violence Against All Women and Girls

Equal mobility means that all groups feel safe regardless of the mode of transport used. This means that the transportation system should be constructed in such a way that both women and men can utilize all parts of the system (including transit environments, parking, streets, cycle paths and pavements) at all times of the day, to maximize mobility. In addition, this goal includes bodily integrity for both women and men.

A possible operationalization of this goal could be

> Risk and fear of exposure to gendered violence or crime in relation to transportation to education and work, as well as the negative impacts of gendered violence on mobility, should be eliminated.

Most gender integration efforts that have been made so far in transport planning are associated with this goal. To ensure safe travel, there should be good lighting, ample open space and monitoring technology (e.g., CCTV) or guards in metro cars and at metro stations. Furthermore, sexual exploitation should be avoided in advertisements and billboards near transport infrastructure. Here, the objective must be adapted to the space in vehicles *and* in environments around transit areas, through applying a 'whole journey' approach. By this, we mean that transport infrastructure includes all steps of a journey, including the space outside residential houses, parking areas of shopping malls, public transport modes and bus terminals.

Box 5.1 Democracy and Influence on Decision-Making: Examples of Targets and Indicators Used in the Analysis
Planning/project organization: How many women and men are involved in the project (50/50% or 40/60% of both sexes does not automatically mean equality)? Who can influence what decisions? How many men and women are in management positions? Who is communicating the plan or project to the affected public, authorities and organizations?

Consulting parties: How many women and men participate in public meetings? How many women and men, from both the public and the competent authorities, are actively involved in the consultation process? How are the men and women heard? For example, how do the experts answer questions and how do they take into account the views of women and men? What questions are answered more extensively?

Based on these questions, targets and indicators can be formulated and applied in the planning process. If the impact assessment finds large differences between men and women, the planning team needs to address these inequalities in the planning process or suggest measures to reduce or compensate for inequalities later in the process.

Box 5.2 Eliminate All Forms of Violence Against All Women and Girls: Examples of Targets and Indicators Used in the Analysis
Number of women and men from various groups who perceive the affected area as safe and secure. Number of women and men from various groups who use the area at different times of day. Examples of groups concerned are children, youth, young adults, adults, older adults, the elderly, those of various ethnic, religious and socioeconomic backgrounds, and transgender or LGBTQIA (lesbian, gay, bisexual, transsexual, queer, intersex and asexual) individuals.[3]

Various methods can be used to analyse the consequences of a plan or programme with regard to this goal. Methods for involving various groups and acquiring knowledge of their experiences include interviews, focus groups, surveys and web-based tools for dialogue with citizens, walking tours, ethnographic fieldwork, thematic consultations and identifying spokespersons. If indicator analysis finds large differences between men and women, the planning team should address these inequalities within the planning process.

Equal Opportunities for Good Health and Personal Development, Including Access to Culture and Leisure

Health and personal development are important to everyday life, and are linked to aspects of daily mobility (cf. Law 1999). Women and men, girls and boys should have equal rights to and opportunities for personal development, education, leisure activities and social networks. Transportation infrastructure is part of society and the environment of people's everyday lives. At the same time, the transportation system has an impact on health by means of traffic accidents, air pollution and noise. However, 'healthy transport planning' can help reduce these risks, as well as promoting walking and cycling.

A possible operationalization of this goal could be

The transport system should minimize the risk of injuries and of poor health due to pollution (i.e., noise and air pollution) for women and men, girls and boys.
The transport system should give women and men, girls and boys the same access to leisure activities.

Health has many definitions, and the determinants of health are factors, including gender, that influence health status and determine health differentials or inequalities (Labonté 1994; Lalonde 1974). In the GIA of transport plans, the assessment of this goal relates to the changes in health risk or personal development of women and men that are reasonably attributable to the specific plan or project.

In practice, this means that planners should ask questions such as: Does the plan/project reduce the impact of air pollution and noise on sensitive groups (e.g., children, the elderly and people with health concerns)? Does the plan/project help ensure that all groups (e.g., women/men, boys/girls) have the same opportunities for physical activity and mobility? These questions can be answered by considering, for example, gender contracts, the numbers of women and men of all age groups travelling in a given area, and air quality and noise models.

Box 5.3 Equal Opportunities for Good Health and Personal Development, Including Access to Culture and Leisure: Examples of Targets and Indicators Used in the Analysis
Number of schools and leisure activities reachable by bike, walking or public transport that children themselves can use. The ability to make 'combination trips' in which several errands can be done on the way to and from work. 'Whole journey' approach and multi-modal trips: combination of walking/cycling, public transport, car-pools, car-sharing and so on.
Access to public transport in connection with primary schools, high schools, colleges and universities.

Equal Opportunities and Conditions for Education and Paid Work

If the transport system is seen as facilitating the achievement of gender objectives, then its design is central to upholding various parts of daily life for women and men. A possible operationalization of this goal could be:

The infrastructure system should contribute to a society in which women and men have the same access to education and paid work, which will provide them with the means to achieve lifelong economic independence.

This and the goal concerning unpaid housework are the two goals for which the transport system has the greatest potential to contribute to the overall gender equality goals of a country. Aspects that can be analysed in assessing this goal are the employment rates of men and women, the educational levels of women and men, and the market regions accessible to women and men. Furthermore, the impact of available public transport to workplaces and geographical patterns of commuting for women and men can be analysed. Are there several possible transport modes, and how is time distributed for different groups of travellers? Is there a possibility of making 'combination trips', in which several errands can be performed on the way to and from work? This goal can be evaluated using gender contracts, interviews, focus groups, statistical analysis and questionnaires.

Equal Distribution of Unpaid Housework and Provision of Care

As in the case of the previous goal, the design of the transport system is central to facilitating various aspects of the daily lives of women and men, including unpaid housework and care provision. Based on this consideration, a possible operationalization of this goal could be

> The infrastructure system should help create conditions that permit equal responsibility for housework for both women and men.

This means that it is important to analyse and assess the access to various transport modes of different groups of travellers. In addition, the ability to make combination trips in which several errands can be performed on the way to and from work is central to evaluating this aspect. The linkages between different destinations, such as schools, workplaces, service centres and sports venues, can be analysed and evaluated. This goal can be evaluated using gender contracts, interviews, focus groups, statistical analysis and questionnaires.

Box 5.4 Equal Opportunities and Conditions for Education and Paid Work: Examples of Targets and Indicators Used in the Analysis
Number of female- and male-dominated workplaces reachable within a certain time (e.g., 15, 30, 40 and 60 min). Gender contracts and how women and men are travelling in the area. Number of schools and leisure activities reachable by bike, walking or public transport that children themselves can use.

Box 5.5 Equal Distribution of Unpaid Housework and Provision of Care: Examples of Targets and Indicators Used in the Analysis
Employment rates and educational levels of men and women in the area affected, labour market regions available to women and men, and the accessibility of these regions by public transport and multimodal travel. Furthermore, geographical directions of commuting for women and men and the number of female- and male-dominated workplaces that can be reached within a certain time (e.g., 15, 30, 40, and 60 min) can be studied. Gender contracts and how women and men travel in the area. Number of schools and leisure activities reachable by bike, walking or public transport that children themselves can use.

Discussion

Aspects of society, including the transport sector, that want to support and achieve gender equality often construe it in terms of gender mainstreaming. This entails ensuring that gender perspectives and attention to the goals of gender equality are central to all activities, such as policy development, research, advocacy, dialogue, legislation, resource allocation, planning and the implementation and monitoring of programmes and projects.

However, the mainstreaming strategy has often been too weak and has proven not to work very well (e.g., Connell and Wood 2005; Forsberg and Lindgren 2015). The GIA model offers a more systematic approach to integrating gender into transport planning processes. The model is based on the results of several research projects on gender mainstreaming in transport planning in a Swedish context. In this chapter, we have added the UN's Sustainable Development Goals on Gender Equality (SDG5) to the model. At the same time, it is important to recognize that the GIA model has been greatly influenced by internationally established IA methodologies, especially SIA and SEA. SIA and SEA were selected as bases for the model because, for example, environmental impact assessments (EIAs) are mandatory in Sweden for many transportation and land-use plans. This means that transport planners are familiar with the IA process. By applying a GIA model that originates from IA, we believe that it will be easier to integrate gender equality into transport planning.

One important aspect of the GIA model presented here is that it differs from the model developed by the EU, also called GIA (cf. European Commission 1997). The main difference is that our model uses gender policy goals as a basis for the objectives used in determining the impacts of a transport plan or project, while the EU model focuses on evaluating how well gender considerations are integrated into new policies. Our criticism is that the EU model constitutes an *ex post* evaluation of gender mainstreaming, meaning that it gives limited support to transport planners at the project level. The aim of our model is to apply an *ex ante* approach to integrating gender into transport planning. In addition, the model is intended to rationalize the decision-making process by giving decision-makers a more complete basis for their decisions.

The GIA model offers a way to apply gender equality goals and objectives formulated in policies applicable to transport planning. The relevant policies may be international (i.e., SDG 5) or national (e.g., Sweden's gender equality objectives), or even objectives formulated at the regional and municipal levels.

An objectives-led IA approach predicts whether or not a proposed action will help fulfil a range of objectives. In this, the strategic assessment acts as an independent 'sustainability/environmental yardstick' against which a strategic action can be measured (Thérivel 2010). In the objectives-led GIA model, the integrative work on gender equality can primarily be linked to the overarching goal of *ending all forms of discrimination against all women and girls everywhere*, and then operationalized as objectives linked to each of the five categories of policy goals:

A. Eliminate all forms of violence against all women and girls
B. Democracy and influence on decision-making
C. Equal opportunities for good health and personal development, including access to culture and leisure
D. Equal opportunities and conditions for education and paid work
E. Equal distribution of unpaid housework and provision of care.

For each of these policy-goal categories, we have formulated objectives, targets, and indicators that can be used as tools in development and planning practice. For each goal category, several objectives can be formulated depending on how the strategic action is expressed. For example, the objective of democracy and influence on decision-making (category B above) can be assigned targets in terms of the proportions of women and men involved in the project: 50/50% or 40/60%. This may be a good target, although equal numbers of women and men do not automatically mean equality. One also needs to ask questions such as: Who can influence what decisions? How many men and women are in management positions? Who is communicating the plan or project to the public, authorities, and organizations affected by it?

As we recognize that the categorizing of policy goals and objectives can be normative, it is important to keep the categories open to reinterpretation. Given that such categorizations always end up sorting people into pre-defined groups, it is important to keep the categories open, to see the potential for reinterpretation and for variation within and between categories (e.g., people who choose not to define themselves as male or female). Our method is intended not to be static but dynamic, in line with the need to understand the complex reality to be operationalized (already, the choice of tools for the operationalization procedure can be problematized). We argue that the suggested operationalization method is not a complete solution for gender mainstreaming, which is an ongoing process. This means that a more equal transport system will be achieved by continually applying models like this one over the long term. We also want to draw attention to situations in which the categories of male/female/LGBTQIA are irrelevant to the problems to be solved—perhaps, in some situations, other social aspects stand out even more clearly in combination with sex and gender. Our view is that both the UN policy and the Swedish national policy need to apply a more intersectional approach. Hanson (2010) noted the importance of being aware that poststructuralism views gender as a socially constructed system of dynamic difference, coexisting with the still very powerful view that gender is an essential source of "fixed and universal female/male difference" (Hanson 2010: 8). Much previous research has found that the transportation system is dominated by technology experts (cf. Mellström 2002, 2017) and that male values and hegemony can hinder the achievement of gender equality goals (cf. Balkmar 2012, and in this volume; Forsberg and Lindgren 2015).

How can the employees (e.g., civil servants and transport planners) of an organization help to realize the gender equality objectives? For professionals, who are confronted every day with various difficult-to-solve questions (cf. Callerstig and Lindholm 2011), conflicting objectives or dilemmas in their work, it is a matter of competence and planning methodology (cf. Henriksson, this volume). We suggest using experts who have knowledge about gender equality and impact assessment. Our experience is that professionals who have previous experience of systematic gender equality work conduct better GIAs. Both women and men can have that experience and should be involved in transportation and infrastructure planning. Planning is a political process that entails negotiation between social forces in a struggle for power. In practice, planning is both political- and value-laden.

Planning is also a matter of management. In previous research, we have suggested that management needs to be involved and take responsibility (Faith-Ell et al. 2010; Faith-Ell and Levin 2012; Levin and Faith-Ell 2011). If gender equality (or for that matter, equality in general) is to permeate policy, it will require political will. When governance fails, or is defined in overly vague terms, not much happens, at least not at a comprehensive level. Svedberg (2014) shows that, without clear instructions from politics, gender goals are rarely implemented in practice. Transport planners, actively working on gender issues and challenging stereotypical norms and beliefs in their daily practice, can bring about changes benefitting women and men, girls and boys, as well as those who do not want to define themselves according to these categories (cf. the chapters by Balkmar, Joelsson, Henriksson, and Rømer Christensen, in this volume). However, to achieve greater social change, clear political leadership is needed. The further out in an organization the responsibility for gender mainstreaming ends up, the more unsystematic it becomes (which is not always negative). It also tends to be tentative about the extent to which citizens may have access to equality work. To ensure comprehensiveness and quality, political work needs to be integrated into the politics, that is, institutions. Research on public administration has found that contradictions can be problematic if they are ignored, but productive if used as a starting point for democratic discussions. Anne-Charlotte Callerstig and Kristina Lindholm's (2011) case studies of gender mainstreaming have described the development of experience in bureaucracy in terms of three phases: problem orientation, examination, and analysis and reflection. Moreover, for GIA to be successful, follow-up is needed. The International Association for Impact Assessment (IAIA) defines EIA followup as the monitoring, evaluation, management and communication of the environmental performance of a project or plan (IAIA 2016; Morrison-Saunders et al. 2007). The same applies to GIA. By incorporating feedback into the GIA process, follow-up enables learning from experience to occur. Follow-up should be applied as a part of every GIA process to prevent GIA from becoming merely a pro forma exercise in transport planning.

IA instruments are often characterized as neutral tools rather than components of the policy process. However, IA instruments also play a role in contemporary political culture (and hence democracy) in governance contexts. "What is less well recognised is that not only are they components of political systems, but impact assessment instruments themselves are also highly politicised, overtly and often covertly" (Cashmore et al. 2010: 178).

If we, like Chantal Mouffe (2005, 2013), choose to see the division between politics and the political as our theoretical basis, we can assume that gender equality work needs to be much more thoroughly examined in politics. Politics concerns institutions (e.g., parties, parliaments and public administrations), while the political is all about battles, conflict, negotiation and distribution issues. In many cases, this is also a matter of monitoring to ensure that directives are followed up, such as the planning of essential transport services, and of understanding that planning dynamics consist of managing conflicting interests and perspectives. We believe that our model provides a useful toolkit and an applicable procedure. Our research has drawn attention to the shortcomings of gender mainstreaming in transport planning as a technology-dominated field of knowledge. This means that we believe that management in transport planning needs to be clearer about the need for gender mainstreaming and that there is a need to prioritize the issue at a political level. This applies even to a country such as Sweden where gender equality has long been on the political agenda, but also internationally in the UN context where women's participation in transport planning has historically been low and gender equality has been considered a women's issue. What we see now, with the introduction of the sustainability goals, is a change in words. Let us hope for action as well.

CONCLUSIONS

The political level is responsible for the development of gender equality policy goals and of systems supporting the realization of policy goals in practice. This highlights mobility as an important aspect and the transport system as a facilitator (i.e., enabler) of the development of equality in the organization of everyday life.

By introducing the objectives-led GIA model, we believe that gender policy goals can be integrated into transport planning.

An objectives-led approach to GIA would aid transport planning with gender mainstreaming. GIA could *improve the outcomes of transport planning*, by providing a systematic tool for integrating gender equality goals. GIA would also *provide an assessment* of whether or not various strategic actions are successful. This means that transport planners would be able to evaluate the *fulfilment of the 2030 Agenda goals* or national gender equality goals. The proposed model could *establish assessment criteria to be used for testing the effects on gender equality* of strategic actions within transport plans. And, finally, GIA based on objectives would also *highlight potential goal conflicts*.

Furthermore, since GIA has the same procedural steps as, for example, EIAs and SEAs, which are mandatory for most Swedish transport and municipal plans, we believe that GIA will have more impact on planning than many other previously suggested models.

Addressing gender equality in spatial planning entails, for example, seeing and understanding why there are gender differences in the public domain and working to eliminate inequalities based on these differences. Given that certain experts (e.g., environmental and economic) are routinely included in transport infrastructure planning, we suggest that gender experts should also be routinely included in transport planning teams (cf. Faith-Ell and Levin 2013; Levin and Faith-Ell 2011).

Acknowledgement This chapter was supported by the Swedish Governmental Agency for Innovation Systems VINNOVA, Grant #2013-02700.

NOTES

1. Transport Analysis is a government agency that provides decision-makers in the sphere of transport policy with evaluations and policy advice.
2. BRÅ primarily works to reduce crime and improve levels of safety in society by producing data and disseminating knowledge about crime and crime-prevention work (http://www.bra.se).
3. LGBTQIA is an inclusive acronym for people with non-mainstream sexual orientations or gender identities.

REFERENCES

Alnebratt, K. (2009). *Meningen med genusforskningen – så som den framträder i forskningspolitiska texter 1970–2000.* [The meaning of gender studies as it appears in research policy documents 1970–2000]. PhD dissertation, Gothenburg University, Sweden.

Alnebratt, K., & Rönnblom, M. (2016). *Feminism som byråkrati* [Feminism as bureaucracy]. Stockholm: Leopard förlag.

Balkmar, D. (2012). *On men and cars: An ethnographic study of gendered, risky and dangerous relations.* PhD dissertation, Linköping University, Sweden.

Callerstig, A.-C., & Lindholm, K. (2011). Det motsägelsefulla arbetet med jäm-ställdhetsintegrering. [The contradictory work on gender mainstreaming]. *Tidskrift för genusvetenskap TGV, 2–3,* 79–96.

Cashmore, M., Richardson, T., Hilding-Ryedvik, T., & Emmelin, L. (2010). Evaluating the effectiveness of impact assessment instruments: Theorising the nature and implications of their political constitution. *Environmental Impact Assessment Review, 30,* 371–379.

CEMR. (2006). *The European charter for equality of women and men in local life.* Innsbruck May 2006.

Christensen, H. R., Poulsen, H., Oldrup, H. H., Malthesen, T., Breengaard, M. H., & Holmen, M. (2007). *Gender mainstreaming European transport research and policies: Building the knowledge base and mapping good practices.* Copenhagen: TRANSGEN.

Connell, R. W. (1996). *Masculinities.* Sydney: Allen and Unwin.

Connell, R. W. (2006). *Om genus.* [About Gender]. Göteborg: Daidalos.

Connell, R. W., & Wood, J. (2005). Globalization and business masculinities. *Men and Masculinities, 7*(4), 347–364.

Dymén, C., Hiselius, L., Kronsell, A., & Smidfelt Rosqvist, L. (2017). Gender equality and increased energy efficiency in the transport sector. Paper for the Workshop *Gendering Smart Mobilities in the Nordic Region,* Oslo 24–25 August 2017.

EIGE. (2017). *Gender equality index 2017: Progress at a snail's pace*. European Institute for Gender Equality. Retrieved November 19, 2017, from http:// eige.europa.eu/

European Commission. (1997). *A guide to gender impact assessment*. Directorate-General for Employment, Industrial Relations and Social Affairs Unit V/D.5. Luxembourg.

European Commission. (2015). *Strategic engagement for gender equality 2016–2019*. Luxembourg: Publications Office of the European Union. Retrieved November 19, 2017, from https://ec.europa.eu/anti-trafficking/eu-policy/strategic-engagement-gender-equality-2016-2019_en

Eveline, J., Bacchi, C., & Binns, J. (2009). Gender mainstreaming versus diversity mainstreaming: Methodology as emancipatory politics. *Gender, Work and Organization, 16*(2), 198–216.

Fainstein, S., & Servon, L. J. (Eds.). (2005). *Gender and planning*. New Brunswick/New York/London: Rutgers University Press.

Faith-Ell, C., & Levin, L. (2012). Jämställdhet och genus i infrastrukturplanering – en studie av tillämpningen inom järnvägsplaneringen. [Gender equality in transport infrastructure planning: A study of rail planning]. Stockholm and Linköping: VTI Report 768. Retrieved April 12, 2017, from https://www.vti.se/sv/Publikationer/Publikation/jamstalldhet-och-genus-i-infrastrukturplanering_670647

Faith-Ell, C., & Levin, L. (2013). *Kön i trafiken. Jämställdhet i kommunal transportplanering*. [Gender mainstreaming in transport planning: Guidance for regional and local transport planners.] Stockholm: SKL [The Swedish Association of Local Authorities and Regions, SALAR]. Retrieved April 1, 2016, from http://webbutik.skl.se/sv/artiklar/samhallsbyggnad/kon-i-trafiken-jamstalldhet-i-kommunal-transportplanering.html

Faith-Ell, C., Levin, L., Dahl, E., Engelbrektsson, E., Nilsson, S., & Yazar, M. (2010). Jämställdhet i samrådsprocesser vid svenska vägprojekt. Genusperspektiv på annonsering, deltagande och mötesinteraktion vid samråd med allmänheten [Gender equality of public participation in Swedish road building projects: A gender perspective on announcements, participation and interaction in public meetings]. Linköping and Stockholm: VTI Report 700.

Forsberg, G., & Lindgren, G. (2015). Regional policy, social networks and informal structures. *European Urban and Regional Studies, 22*(4), 368–382.

Gil Solá, A. (2016). Constructing work travel inequalities: The role of household gender contracts. *Journal of Transport Geography, 53*, 32–40.

Goffman, E. (1971). *Relations in public: Microstudies of the public order*. New York: Basic Books.

Goffman, E. (2004). *The presentation of self in everyday life* (4th ed.). Stockholm: Norstedts akademiska förlag.

Government Bill 2005/06:155. Makt att forma samhället och sitt eget liv – nya mål i jämställdhetspolitiken. [Power to shape society and their own lives: New gender policy objectives]. Government Offices of Sweden. (2016). Mål för jämställdhet. [Objectives for gender equality]. Retrieved September 21, 2017, from http://www.regeringen. se/regeringens-politik/jamstalldhet/mal-for-jamstalldhet/

Göransson, A. (Ed.). (2007). Maktens kön. Kvinnor och män i den svenska makteliten på 2000-talet. [The gender of power. Women and men in the Swedish power elite in the 2000s]. Nora: Nya doxa.

Halling, J., Faith-Ell, C., & Levin, L. 2016. Transportplanering i förändring: En handbok om jämställdhetskonsekvensbedömning i transportplaneringen. [Transport planning in change: A handbook on gender impact assessment]. Linköping/ Lund/Stockholm: K2. Retrieved December 1, 2016, from http:// www.k2centrum.se/transportplanering-i-forandring-en-handbok-om-jamstalldhetskonsekvensbedomning

Hanson, S. (2010). Gender and mobility: New approaches for informing sustainability. Gender, Place and Culture, 17(1), 5–23.

Hirdman, Y. (1993). Genussystemet – reflektioner om kvinnors sociala underordning [The gender system: Reflections on the subordination of women]. In C. Ericsson (Ed.), Genus i histoisk forskning [Gender in historical research]. Lund: Studentlitteratur.

Hirdman, Y. (2003). Genus: Om det stabilas föränderliga former [Gender: On the shifting forms of stability]. Malmö: Liber.

IAIA Social Impact Assessment. Retrieved April 1, 2016, from http://www.iaia. org/wiki-details.php?ID=23

Jayyusi, L. (1984). Categorization and the moral order. Boston: Routledge and Kegan Paul.

Joelsson, T. (2013). Space and sensibility: Young men's risk-taking with motor vehicles. PhD dissertation, Linköping University, Sweden.

Kronsell, A. (2005). Gendered practice in institutions of hegemonic masculinity: Reflections from feminist standpoint theory. International Feminist Journal of Politics, 7(2), 280–298.

Labonté, R. (1994). Health promotion and empowerment practice frameworks: Issues in health promotion services HP-10-0102. Saskatchewan Population Health and Evaluation Research Unit, Canada.

Lalonde, M. (1974). A new perspective of the health of the Canadians: A working document. Government of Canada, Ottawa, April 1974.

Law, R. (1999). Beyond 'women and transport': Towards new geographies of gender and daily mobility. Progress in Human Geography, 23(4), 567–588.

Levin, L., & Faith-Ell, C. (2011). Genusperspektiv på utveckling av kollektivtrafik. [Gender perspective on the development of public transport]. Linköping and Stockholm: VTI Report 712. Retrieved April 12, 2017, from https://www.vti. se/sv/Publikationer/Publikation/genusperspektiv-pa-utveckling-av-kollektivtrafik_670584

Levin, L., & Faith-Ell, C. (2014). Methods and tools for gender mainstreaming in Swedish transport planning. Proceedings from *The 5th international conference on women's issues in transportation – Bridging the gap. 14–16 April 2014, Paris – Marne-la-Vallée, France*. TRB, pp. 215–223.

Lombardo, E., & Meier, P. (2008). Framing gender equality in the European Union discourse. *Social Politics, 15*(1), 101–129.

Lombardo, E., Meier, P., & Verloo, M. (2010). Discursive dynamics in gender equality politics: What about 'feminist taboos'? *European Journal of Women's Studies, 17*(2), 105–123.

Mellström, U. (2002). Patriarchal machines and masculine embodiment. *Science Technology, & Human Values, 27*(4), 460–478.

Mellström, U. (2017). *Masculinity, power and technology*. London: Routledge.

Morrison-Saunders, A., Marshall, R., & Arts, J. (2007). *EIA follow-up international best practice principles*. Special Publication Series No. 6. Fargo, USA: International Association for Impact Assessment.

Mouffe, C. (2005). *The return of the political*. London: Verso Books.

Mouffe, C. (2013). *Agonistics: Thinking the world politically*. London: Verso Books.

Sadler, B., & Verheem, R. (1996). *SEA: Status, challenges and future directions*. Report 53, Ministry of Housing, Spatial Planning and Development (VROM), The Hague Netherlands.

Sandercock, L., & Forsyth, A. (2005). A gender agenda: New directions for planning theory. In S. Fainstein & L. J. Servon (Eds.), *Gender and planning* (pp. 67–85). New Brunswick/New York/London: Rutgers University Press.

Sharp, J., Briggs, J., Yacoub, H., & Hamed, N. (2003). Doing gender and development: Understanding empowerment and local gender relations. *Transactions of the Institute of British Geographers, 28*, 281–295.

Svedberg, W. (2014). *Ett (O)jämställt transportsystem i gränslandet mellan politik och rätt – En genusrättsvetenskaplig studie av rättslig styrning för jämställdhet inom vissa samhällsområden*. [A Gender (Un)Equal Transport System in the borderland between Policy and Law – A Gender Legal Study of legal governance for Gender Equality in certain areas of society]. PhD dissertation, Gothenburg University, Sweden.

Therivel, R. (2010). *Strategic environmental assessment in action*. London: Earthscan.

Tienari, J., Meriläinen, S., Holgersson, C., & Bendl, R. (2013). And then there are none: On the exclusion of women in processes of executive search. *Gender in Management: An International Journal, 28*(1), 43–62.

Trafikanalys. (2017). Uppföljning av de transportpolitiska målen 2017 [Follow-up of the transport policy goals 2017]. Transport Analysis. Report 2017:7.

UN. (2015a). *Sustainable development goals: A SRHR CSO guide for national implementation*. London: International Planned Parenthood Federation (IPPF). Retrieved April 12, 2017, from http://www.ippf.org/sites/default/files/sdg_a_srhr_guide_to_national_implementation_english_web.pdf

UN. (2015b). *Transforming our world: The 2030 agenda for sustainable develop-ment.* Retrieved December 1, 2016, from https://sustainabledevelopment. un.org/post2015/transformingourworld

Vanclay, F. (2003). International principles for social impact assessment. *Impact Assessment and Project Appraisal, 21*(1), 5–11.

West, C., & Zimmerman, D. H. (1987). Doing gender. *Gender and Society, 1*(2), 125–151.

Woodward, A. (2003). European gender mainstreaming: Promises and pitfalls of transformative policy. *Review of Policy Research, 20*(1), 65–88.

PART III

Gendering Travel Surveys

CHAPTER 6

Til Work Do Us Part: The Social Fallacy of Long-Distance Commuting

Erika Sandow

INTRODUCTION

Commuting is an important aspect of daily life for many people. The majority of the workforce travels to and from work on a daily basis. Presently in Europe, there is an increasing trend of long-distance commuting, both daily and weekly (Barker and Connolly 2006; Lyons and Chatterjee 2008; Renkow and Hoover 2000).

This development of increased commuting over longer distances is likely to continue in the coming years and will have consequences on both societies and individuals. The environmental effects of commuting, such as air pollution, urban sprawl and traffic congestion, are well known (Brueckner 2000; Tolley 1996; Travisi et al. 2010), as is the importance of commuting for one's economic performance on the labour market (Krieger and Fernandez 2006; Russo et al. 2007). That commuting longer distances is

This article is a reprint from 2014 in the journal *Urban Studies*, 51(3), 526–543.

E. Sandow (✉)
Department of Geography and Economic History/Centre for Demographic and Ageing Research CEDAR, Umeå University, Umeå, Sweden
e-mail: erika.sandow@umu.se

© The Author(s) 2019
C. L. Scholten, T. Joelsson (eds.), *Integrating Gender into Transport Planning*, https://doi.org/10.1007/978-3-030-05042-9_6

121

often associated with higher incomes and career opportunities for the individual is also acknowledged (So et al. 2001; van Ommeren et al. 2000).

Long-distance commuting also imposes significant costs on people and their social environment (Cassidy 1992; Green et al. 1999). Regardless of the individuals' motives, commuters must often spend less time socializing with family and friends as they are away from home a longer time, either daily or several days a week if they commute weekly. It is not always unproblematic for couples to manage to balance work and everyday life in a sustainable lifestyle when one or both of them commute long distances on a daily basis. The picture of the (long-term) social implications of long-distance commuting on commuters and their households is relatively unclear, however. The main reason for this is that many studies have mainly focused on either the personal well-being of the commuter himself or herself (e.g. health, stress), or economic aspects regarding commuting, often in terms of income and career achievements or when commuting is an alternative to migration. While there are several studies that focus on the increase in commuter marriages in which couples live apart during the week (Green et al. 1999; van der Klis and Karsten 2009; van der Klis and Mulder 2008) there are few, if any, longitudinal studies that assess the long-term effects of commuting on family relationships in terms of divorce/separation. Therefore, this article focuses on the behaviour of couples when one or both spouses are long-distance commuting (30 km or more). The aim is to analyse how long-distance commuting affects the risk of separation among Swedish couples.

The chapter utilizes a unique dataset with longitudinal data. This enables the empirical analysis of the effect of long-distance commuting on couples, which can hardly be done using questionnaire or interview surveys. Discrete-time logistic regression models were employed to register data on Swedish couples (two adults, with or without children) in 2000 and their commuting behaviour over a ten-year period (1995–2005) to analyse whether long-distance commuting appears to affect the duration of marriages/cohabitations ending in divorce or separation.

The rest of this article is organized as follows: the next section gives a brief overview of literature regarding social implications of long-distance commuting on people and households. Following this section, the data as well as the regression methods used to estimate the risk of separation are described. In the fourth section, some descriptive results and estimation results are presented and interpreted. These results are finally discussed in the concluding section of the article.

THEORETICAL FRAMEWORK: HOW DOES COMMUTING AFFECT COMMUTERS AND THEIR HOUSEHOLDS?

From a classical urban economic perspective, extra monetary and mental costs of commuting longer distances must be compensated for on the labour and housing markets in order for the commuters' well-being or utility to be equalized (Alonso 1964; Muth 1969). For example, a long commuting distance can be compensated for by lower rents or housing prices (e.g. Renkow and Hoover 2000), desired housing or neighbourhood characteristics (e.g. Plaut 2006) or an intrinsically or financially rewarding job (e.g. So et al. 2001; van Ommeren et al. 2000; Zax 1991). Accordingly, commuting is then determined to be in an equilibrium state on the housing and labour market in which the commuters are fully compensated with either higher wages or lower rents. On average, commuters' subjective well-being or utility would then be the same regardless of their commuting time.

Commuting is more than simply covering a physical distance between home and work. Besides taking time, it also involves travelling costs and can affect individuals' personal well-being in several ways and intervene in personal relationships.

Commuting and Subjective Well-Being

There are studies that show that commuters on average are less satisfied with their lives than non-commuters are. In a study by Stutzer and Frey (2008), commuters' subjective well-being was studied as a proxy for utility to test whether they are fully compensated for the stress they incur. They found that longer commutes systematically decrease life satisfaction. For example, a commuter who travels one hour each way would have to make 40% more in salary to be as "satisfied" with life as a non-commuter is. They therefore conclude that most commuters are not fully compensated for the stress they pay—a so-called commuting paradox. A Swedish study by Fults (2010) also suggests that longer commuting times have a negative effect on people's subjective well-being. Several other studies also indicate that commuters generally experience a stressful lifestyle impacting their own psychological and physical well-being (Evans et al. 2002; Kluger 1998), and that longer commuting time increases commuters' stress levels (Costa et al. 1988; Gottholmseder et al. 2009; Kluger 1998). Besides the travel time, one's control over the commute has an impact on perceived stress. Commuting is more stressful when there is a lack of control and the journey involves unpredictability, such as traffic stocks, the driving behaviour of other road users (for car users) and unreliability of public transport services (Evans et al. 2002; Koslowsky et al. 1996; Kluger 1998). A questionnaire study by Gatersleben and Uzzell (2007) shows that car drivers have the most stressful commutes, but public transit users had the most negative attitudes towards their travel mode, mainly due to boredom. Kluger (1998) found that car commuters with lengthy journeys are likely to be in a negative mood in the evening, implying that commuting stress can spill over into family life and overall well-being in several ways. Having experience commuting, however, makes the commute more predictable and is thus found to lower the stress induced by commuting (Kluger 1998). Analyses by Gottholmseder et al. (2009) of the effect of commuting in Austria on individuals' stress perception found that commuters living in a partnership could handle stress better than single commuters could, but there were no gender differences. On the other hand, Roberts et al. (2009) found that women in the UK were more sensitive to commuting stress than men were, even when considering working hours and occupations and that this stress is a result of the gendered division of their everyday household tasks, including childcare.

While commuting can cause stress and health problems, in several studies Mokhtarian and colleagues (Mokhtarian and Salomon 2001a, b; Mokhtarian et al. 2001; Ory and Mokhtarian 2005) have shown that the time spent between home and work can also be utilized by the commuter for something positive. Beyond the obvious utility of reaching work, there are utilities for activities that can be conducted while commuting, for example, doing work, or social networking through mobile information and communication technologies or simply sleeping or enjoying the scenery. For the conventional commuting worker, the commute can also offer a natural transition between being at work and at home. Time spent in the car or on the bus, train or bike, or otherwise making one's way from work to home can serve as a decompression period for commuters. There are qualitative studies that suggest that commuting might be understood as a gift rather than a curse (Jain and Lyons 2008) and that some people even want to have longer commuting times in order to be able to use the travel time productively (Redmond and Mokhtarian 2001). Those who derive positive utility from commuting are also found to experience the commute as less stressful (Gottholmseder et al. 2009) and experience less disutility of commuting (Lyons and Urry 2005; Ory and Mokhtarian 2005). Nevertheless, when commuting times become too long, the willingness to commute decreases (Sandow and Westin 2010). There are also studies that show that commuters would like to decrease their commuting time, regardless of the mode used (Sandow and Westin 2010; Páez and Whalen 2010; Redmond and Mokhtarian 2001).

Commuting a Mobility Strategy for the Household

Commuting can be a mobility strategy to maintain social relationships and avoid migration. It can give people the opportunity to maintain a social network built up during years of living at the same place, which would be lost if they moved (Fisher and Malmberg 2001). In post-modern society, shifting labour market structures involving increased specialization make it more difficult for both partners in a dual-career household to accommodate both spouses' careers close to home. Commuting for one or both spouses in a household can, therefore, make it possible for each of them to have a career without moving. For some households, these complexities of the changing geographical labour mobility result in a "commuter marriage/partnership" in which both partners pursue their careers while one lives near work part of the time because the commuting distance is too great to travel on a daily basis (Green et al. 1999; van der Klis and Karsten 2009; van der Klis and Mulder 2008).

While commuter marriages can be a solution for dual-career couples as they theoretically present equal opportunities for couples' careers, for workers within the mining, oil and gas industries, this lifestyle has rather been the result of structural changes within these employment sectors. Within these industries, mainly in Australia and Canada, long-distance commuting (also referred to as fly-in commuting) has become the dominant approach to new mining, oil and gas developments (Costa 2004; Houghton 1993; Storey 2001). Rather than developing new settlements with services to accommodate workers and their families, as a means to reduce costs, a great deal of the workers are transported from metropolitan areas to remote sites to work for a certain number of days, after which they return home for a number of days.

In households with children, this mobility strategy with commuter marriages often means that the home-based parent (often the woman) limits his/her work time in order to juggle work and family commitments when the other spouse is away from home. While at least one partner can pursue a career with this geographical strategy, van der Klis and Karsten (2009) found that among commuter families in the Netherlands the commuting parent experiences sacrifices due to missing vital parts of everyday family life.

Commuting is generally a more difficult lifestyle when there are children living at home. When a parent is away from home for longer periods of time during the day, or for several days if the commute is weekly, he/she can feel guilty for missing vital daily parts of the children's development (Rotter et al. 1998). There is also a mental distance to consider when long-distance commuting. Even if access to fast modes of transportation makes it possible to choose to work further from one's home without having to increase travel time, there is a mental distance. If something happens at school or a child gets sick, it is important for parents of small children to know that at least one of them can manage to get home quickly (Friberg et al. 2004).

Commuting and Citizen Participation

Lengthy commutes leaving fewer hours for spare-time activities also have a direct negative impact on people's involvement in community affairs and informal social interaction (Pocock 2003 cited in Flood and Barbato 2005, p. 7; Vilhelmson 2002). Studies on the relationship between commuting and citizen participation in the US suggest that commuting is likely to reduce the time available for political activism. Putnam (2001), for example, argues that there is a negative relationship between commuting and civic engagement. He estimates that for each additional ten minutes commuting a person's involvement in community affairs (such as attending public meetings or church services and chairing committees) is cut by 10%. Lidström (2006), on the other hand, has found that this is not the case in a Swedish context. On the contrary, Lidström found that different forms of citizen participation are actually more common among commuters than non-commuters. But the results also suggest that this positive relationship holds only as long as the commuting time is limited. If the trend of increasing long-distance commuting continues in Sweden, more people will face more obvious time limits for citizen participation.

Overall, the above studies imply that lengthy commuting means less time for interaction and socializing with family, friends and neighbours as well as for engagement in community affairs. While commuter partnership is an exceptional example of how family structures can change, life for many other couples is also affected by long-distance commuting. One question is then, what the social implications of long-distance commuting on relationships are, and whether or not couples can handle the costs of long-distance commuting in the long run.

DATA AND METHOD

Empirically, the study is based on geo-referenced longitudinal individual register data for the entire Swedish population. The database ASTRID contains annually updated information from Statistics Sweden on many individual demographic and socioeconomic attributes including family situation and members, earnings, work, employment and unemployment, support income and coordinates for place of living and work with 100 metres resolution. Information on travel time and travel modes is unfortunately not available, however. The definition of what constitutes a long-distance commute is therefore based on travel distance as a proxy for travel time. What constitutes a long commuting distance is however not clear and unambiguous. In this study a one-way distance (Euclidean distance) of 30 kilometres or more is defined as a long-distance commute. This corresponds to an average of at least 45 minutes by car in Sweden,[1] which has been found to be the threshold for what constitutes an acceptable travel time both nationally in Sweden and internationally, as longer travel times have costs that impact daily life (see van Ommeren 1996; Sandow and Westin 2010; Wachs et al. 1993). This Euclidean distance between the coordinates of home and work do not measure the actual distance, which is about 30% longer (Reneland 1998), but is likely to correspond to about 35–50 kilometres on the ground depending on context (physical infrastructure, transport mode, traffic conditions etc.). As the data is on an annual basis it is not possible to distinguish between daily and weekly long-distance commuting; therefore, both groups are included in the analysis.

In total, the data consisted of all employed individuals aged 20–60 years in 2000. Persons not living with a partner in the year 2000 were excluded. The focus of this study is nuclear household relationships; thus, only couples living together[2] (married or registered as cohabiting[3] and having the same residential coordinates) were accounted for in the analyses. At the time of the study this gives a total of 2,143,256 persons, of whom 186,156 (9%) were long-distance commuters in year 2000. Information on a number of demographic and socioeconomic characteristics of all these individuals in the sample was extracted from the database back to 1995 and up to 2005.

In this nationwide longitudinal study, the extent to which long-distance commuting increases the risk of separation is investigated through event history analysis using discrete-time logistic regression. As Allison (1982) and Singer and Willett (1993) have shown, this is an adequate method for studying the occurrence of an event, such as a separation. The method is based on a person-year dataset, in which each person has multiple records (one for each year), instead of a person-oriented dataset as in standard logistic regression. Also unlike standard logistic regression, which examines the overall (unconditional) probability of an event without taking into account the timing of that event, discrete-time logistic regression models the conditional probability that the event will occur (i.e. the probability of separating at time t_1 given that the relationship has lasted until t_1). More importantly, discrete-time models have the ability to examine the impact of time-varying covariates, such as the type of place of residence or income from year to year, on the outcome of an event. For example, an individual's household composition (i.e. being a mother or a father) or income may change over time and these changes may have an effect on the risk of experiencing the event. All covariates (i.e. independent variables) used here are time-varying. The demographic covariates (Table 6.1) include age and children living at home each year. The socioeconomic covariates include highest attained education level, employment sector and income each year, deflated according to the value in the year 2000 of the Swedish crown and commuting status each year (commuting or not). Employment is defined as having an annual income from work of at least 50,000 SEK in order to exclude those who are not part of the workforce, for example students working on the side. Furthermore, as the geographical structure of the Swedish local labour market is very diverse, from three major metropolitan regions down to small and sparsely populated regions, it was also tested whether different residential contexts have different impacts on separation for commuters and non-commuters. The choice of regions was based on the Swedish Agency for Economic and Regional Growth's (2010) classification of regions into metropolitan regions, urban regions and rural regions. This classification is based on functional labour market regions in which people can live and work without too much time-consuming commuting. These functional regions are then grouped together based on factors important to development and growth (population size, education level, share of businesses and accessibility to employment opportunities) into five groups of regions: metropolitan regions, urban regions, rural regional centres, rural regions—private employment and rural regions—public employment. Here the three smaller regions are grouped into rural regions. Thus, the analysis estimates the probability of separation for each year as a function of the

Table 6.1 Descriptive characteristics of couples, Sweden, 2000 (*N*=2,143,256)

Characteristics	Women		Men	
	Long-distance commuter (n = 188,306)	Non-commuter (n = 868,545)	Long-distance commuter (n = 312,301)	Non-commuter (n = 774,104)
Gender (percentages)	18	82	29	71
Average age	43	44	44	44
Family situation (percentages)				
Children aged 0–6 years[a]	32	27	33	33
Children aged 7–17 years[a]	40	44	45	45
No children	41	41	36	36
Education level (percentages)				
Junior high school (low)	11	15	16	21
High school (medium)	63	67	64	63
University education (high)	26	18	21	17
Income level (percentages)[b]				
Low	48	59	16	24
Medium	32	28	39	45
High	21	13	45	31
Employment sector (percentages)				
Primary or secondary	15	14	40	48
Private service	29	21	38	31
Public service	51	60	19	17
Other	6	5	4	4
Residential region (percentages)				
Metropolitan	46	45	45	45
Urban	37	39	39	39
Rural	17	16	16	16

[a]Can have children in both age groups
[b]Annual income level: Low 50,000–200,000 SEK; Middle 200,000–300,000 SEK; High 300,000 + SEK. A high income corresponds to approximately 31,300 €

above-mentioned covariates. The outcome (event) of the study was divorce for married persons and a change from cohabiting status to living alone for persons registered as cohabiting.

The logistic regression model for the estimates of separation by person *i* in the year *t* is:

$$\log\left[P_{it} / \left(1 - P_{it}\right)\right] = \alpha_t + \beta' x_{it}$$

Where α_t (t=1, 2, …) is the constant term, x_{it} the explanatory variable and β is the logistic regression coefficient.

All persons were followed from the year 2000 until separation, death, death of spouse or the year 2005, and were then removed from the dataset (censored). Only the recorded marriage/cohabitation in 2000 was considered. This could be a second- or higher-order marriage if the persons had separated or been widowed before 2000. Because information on family status only goes back to 1995, we cannot know the duration of those marriages/cohabitations so it was arbitrarily set to one year. Using an indicator (dummy) it was tested whether or not the marriage/partnership had already started in 1995, but the commuting effects were always the same (not shown). Models were also estimated separately for those marrying/cohabiting before and after 1995, but since they gave very similar results they are not shown.

Individuals or households engage in long-distance commuting for a range of different reasons. As the individual's or household's motives for long-distance commuting are not observable in register data, this chapter concentrates on actual commuting behaviour and those who are in fact long-distance commuting. Several specifications of the long-distance commuting variable were used. In one model, a distinction was made between not commuting and commuting each year to explore the overall effect. In another model, distinctions were made between whether the person had been involved in long-distance commuting for five years or more or less than five years, or had not commuted during 1995–2005. All models were estimated separately for women and men.

RESULTS

Descriptives

In total, 9% of the couples long-distance commuted in 2000. About half (49%) of these people were still long-distance commuting five years later. Overall, it was relatively common to commute for a longer period: 34% of the couples studied long-distance commuted at least five years during the period 1995–2005. It was more common among men than women to be a long-distance commuter (Table 6.1).

Table 6.1 shows socioeconomic and geographical characteristics for long-distance commuters and non-commuters divided by gender. Like most people in Sweden, the majority of the long-distance commuting couples live in metropolitan regions. The share of highly educated and high-income earners is greater among long-distance commuters compared to the non-commuters. There are, however, large gender differences between the commuters. The share of highly educated female commuters is greater than that of highly educated male commuters. Despite these educational differences between commuting women and men, only a minority of these women are high-income earners, while most long-distance commuting men have a high income. In line with another study of long-distance commuters in Sweden (Sandow and Westin 2010), this reflects that men benefit economically from long-distance commuting more than women do, although long-distance commuting generally has a positive effect on a person's working income.

While the average age indicates that many couples are middle-aged, it is more common to have children, especially of pre-school age, among commuters than the non-commuting couples. This may imply that children increase place attachment and that long-distance commuting is part of a strategic mobility choice or solution to avoid migration.

Separation Rates

The total number of persons who had separated at the five-year follow-up in 2000 was 236,446 (11%), of whom 110,253 were women and 126,193 men.[4] On average, long-distance commuter couples have a 40% higher risk of separating than do non-commuting couples (14% of the long-distance commuting couples separated compared to 10% of the non-commuting couples). As shown in Fig. 6.1, long-distance commuters' annual separation rates are higher than those of non-commuters, even though the share of separations decreases with time for the studied couples.

As shown by Sandow and Westin (2010), long-distance commuting in Sweden seems to be a long-term mobility choice rather than a short-term one; The majority of long-distance commuters in Sweden commute for five years or more, and a great deal of them commute for more than ten years. It was therefore tested whether separation rates differ between couples in which one or both partners have a long duration of long-distance commuting compared to other couples. It was found that among couples in which the commuter(s) have a long duration (≥5 years) of long-distance commuting, fewer couples separated (11%) compared to those having a short commuting duration (15%). Separation rates were about the same for non-commuting couples (10%) and couples having a long commuting duration (11%). That the risk of separation is lower among the persevering

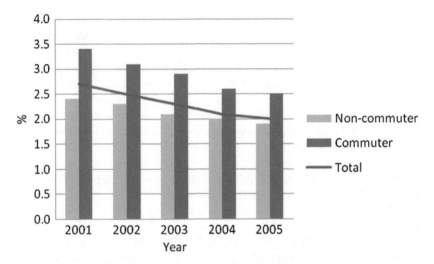

Fig. 6.1 Separation rates, percentage of separations each year among studies couples

commuters may reflect a customisation process (Rüger and Ruppenthal 2010), whereby the many years of long-distance commuting has made the commute part of one's lifestyle, and one has learned to live with the experiences of the social and economic costs and benefits of commuting in everyday life. The gained experiences of what it means to long-distance commute may then be seen as worthwhile. A selection process could then explain that those who lack the ability to manage the stress and other costs caused by long-distance commuting have separated or stopped long-distance commuting to avoid the risk of a separation. Those couples separating during the first years of long-distance commuting may have separated anyway, also reflecting a selection process.

Overall, these differences in separation rates between the two groups of commuters are the same for each year (Fig. 6.2). No gender differences were found regarding who the commuter was in a couple. For those couples who did not separate, it did not matter whether it was the man or woman who was the long-distance commuter.

Previous Experiences of Commuting
About 20% of the long-distance commuters were already commuting when they married or moved in with their partner.[5] For these people, commuting with its pros and cons was probably already part of their every-

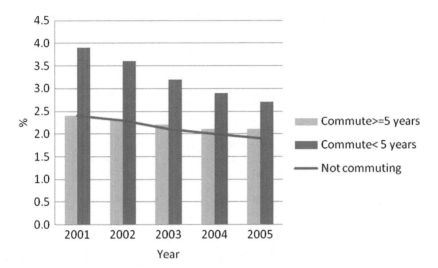

Fig. 6.2   Separation rates divided by years in long-distance commuting, percentage of separations each year among studied couples

day life. When the effect on separation of previous experience of long-distance commuting before moving in together/getting married was controlled for, a positive relationship was found. Among commuting couples, separation was less common if at least one spouse had previous experience of long-distance commuting before marriage/cohabiting, and was long-distance commuting at the time of marriage/cohabitation (16% compared to 20% for those without experience). Separation was least common if it was the female spouse who had experience of long-distance commuting prior to the relationship.

Regression Results

Overall Commuting Effects
Married and cohabiting men who long-distance commuted had 4% higher odds of separating than did men who did not long-distance commute, for whom all the other observed variables were the same (Table 6.2). No overall effect of commuting was seen for women. Estimates with a subdivision of commuting duration (Table 6.3) show higher separation rates among long-distance commuting men only for a duration of less than five years.

Table 6.2   Results from discrete-time logistic regression on the estimation of the effect of long-distance commuting, education, income, children, age and duration of partnership on the odds of separating

	Women	Men
	OR^a (95% CI^b)	OR^a (95% CI^b)
*Age**		
20–24	**0.14 (0.12–0.15)**	**0.13 (0.12–0.14)**
25–29	**0.28 (0.27–0.29)**	**0.23 (0.22–0.24)**
30–34	**0.51 (0.50–0.53)**	**0.40 (0.39–03.42)**
35–39	**0.88 (0.86–0.90)**	**0.78 (0.74–0.78)**
40–44 (ref.)	1 (ref.)*	1 (ref.)*
45–49	**0.90 (0.88–0.92)**	**0.89(0.87–0.91)**
50–54	**0.62 (0.60–0.64)**	**0.55 (0.53–0.56)**
55 +	**0.38 (0.37–0.40)**	**0.31 (0.30–0.32)**
*Children 0–6 years living at home**		
Yes	**0.31 (0.31–0.32)**	**0.06(0.05–0.06)**
No	1 (ref.)*	1 (ref.)*
*Children 7–17 years living at home**		
Yes	**1.11 (1.09–1.13)**	**0.37 (0.37–0.37)**
No	1 (ref.)*	1 (ref.)*
*Education level**		
Low	**1.26 (1.22–1.30)**	**1.44 (1.40–1.48)**
Middle	**1.36 (1.34–1.39)**	**1.48(1.45–1.51)**
High (ref.)	1 (ref.)*	1 (ref.)*
Income level		
Low	**0.86 (0.84–0.88)**	1.01 (0.99–1.03)
Medium	**1.19 (1.16–1.21)**	**1.03 (1.01–1.05)**
High (ref.)	1 (ref)*	1 (ref.)*
*Employment sector**		
Primary and secondary	1 (ref.)*	1 (ref.)*
Private service	**1.10 (1.07–1.13)**	**1.02 (1.00–1.04)**
Public service	**1.13 (1.11–1.16)**	**1.12 (1.10–1.15)**
Other services	**1.09 (1.05–1.13)**	1.03 (0.99–1.07)
Duration partnership (per year)	**0.47 (0.47–0.48)**	**0.49 (0.48–0.49)**
*Long-distance commuter**		
Yes	1.00 (0.98–1.02)	**1.04 (1.02–1.05)**
No	1 (ref.)	1 (ref.)*
	Log likelihood = 606,391,282; pseudo R^2 Nagelkerke= 0.308	Log likelihood = 623,992,339; pseudo R^2 Nagelkerke= 0.367

*Indicates that the whole variable is significant at $p \leq 0.01$, that is, it shows whether or not the model as a whole becomes better when the variable is included in it
[a]Odds ratio, significant values ($p \leq 0.01$) are marked with bold text. Values > 1 show increased odds
[b]95% confidence interval

Table 6.3   Results from discrete-time logistic regression on the estimation of the effects of the duration of long-distance commuting on separating

	Women	Men
	OR[a] (95% CI[b])	OR[a] (95% CI[b])
Long-distance commuter		
No	1 (ref.)*	1 (ref.)*
1–4 years	1.02 (1.00–1.04)	**1.04 (1.02–1.06)**
5 + years	**0.93 (0.90–0.96)**	1.02 (1.00–1.05)
Log likelihood	606,366.877	623,990.564
Pseudo R² Nagelkerke	0.308	0.367

Note: Only the effects of the commuting variable are displayed in this table, but all the variables shown in Table 6.2 were included in the models and the estimates for these variables were the same in both models
*Indicates that the whole variable is significant at $p \leq 0.01$, that is, it shows whether or not the model as a whole becomes better when the variable is included in it
[a]Odds ratio, values > 1 show increased odds. Significant values ($p \leq 0.01$) are marked with bold text
[b]95% confidence interval

Men long-distance commuting over a longer time period do not separate to a significantly higher extent than non-commuting men do. In contrast to the male long-distance commuters, a long duration of long-distance commuting for women is associated with an 8% reduction in the odds of separating (Table 6.3).

Models (results not shown here) were also run to test the effect of long-distance commuting on relationships when both partners were involved in long-distance commuting as well as the effect of previous experiences of long-distance commuting. If both spouses were long-distance commuting, separation rates were lower. Having previous experience of long-distance commuting when starting a relationship was also found to reduce separation rates for both women (16%) and men (11%).

While long-distance commuting affects the odds of separation, other factors also have a significant effect on the probability of separating. For example, when other factors are controlled for, the longer the duration of marriage/cohabitation the lower the odds are of separating for both women and men. Also, those living in a family with small children (0–6 years) at home, when other factors are controlled for, separate to a lower extent than do couples with no small children at home.

Children

Couples with children are generally found to be more stable than childless couples, and the lowest risk of separation is seen in couples with very young children (Andersson 1997; Hoem 1997). As a greater share of the long-distance commuters than the non-commuters have pre-school children, it was tested whether the effects of long-distance commuting on separation differ between childless couples and those with small children (0–6 years) living at home. The model was estimated separately for couples with small children in the household and for couples having no children. The results (not shown here) confirm that long-distance commuting increases the risk of separating for all groups except for women with small children; they actually lower their risk of separating when long-distance commuting over a longer time period.

Geographical Effects
The long-distance commuting effect on separation varies in different geographical contexts (see Table 6.4). For example, men living in metropolitan regions have 4–5% higher odds of separating if they are long-distance commuting compared to men who are not commuting (but who have the same income level, duration of marriage, etc.). The highest separation rates for both women and men were estimated for short durations of long-distance commuting in rural regions. After five years of commuting, the divorce rates were equal to those of the general population of men and women living in rural regions. For women living in urban regions, separation rates are 11% lower among persistent long-distance commuters (>5 years) than among non-commuters. The point estimates suggested low divorce rates for women having a long duration of long-distance commuting in the other two residential regions, but significance was not attained.

It is not possible to explain within this study why a few years of long-distance commuting significantly increases the risk of separating for those living in rural regions. Overall, because the reasons these people start (or stop) long-distance commuting—as well as transportation mode—is unknown in this study, it is not possible to say why these geographical differences exist in how long-distance commuting affects household relationships. It could be the case that the reasons people start long-distance commuting on average differ depending on where they live, and that this has an effect on how couples can handle the consequences of long-distance commuting. For example, when someone lives in a smaller region his/her employment opportunities in the local labour market are more limited compared to those in big city regions. The reason to start commuting can, therefore, be more of a necessity in order to acquire a job and avoid migration. The social costs of long-distance commuting can, therefore, be expe-

Table 6.4 Effect of long-distance commuting (A) and the duration of long-distance commuting (B) on the odds of separating by geographical area of residence

	Women			Men		
	OR[a] (95%CI[b])			OR[a] (95%CI[b])		
Residential region	Metropolitan	Urban	Rural	Metropolitan	Urban	Rural
A Long-distance commuter						
Yes	0.98 (0.96–1.01)	0.99 (0.96–1.02)	1.05 (1.01–1.10)	**1.04 (1.02–1.07)**	1.00 (0.97–1.02)	**1.05 (1.01–1.10)**
No	1 (ref.)	1 (ref.)*	1 (ref.)*	**1 (ref.)***	1 (ref.)	**1 (ref.)***
Log likelihood	298,770,03	220,075,68	86,451,814	299,366,12	232,001,03	91,252,662
Pseudo R² Nagelkerke	0.297	0.316	0.326	0.352	0.381	0.385
B Long-distance commuter						
No	1(ref.)	1(ref.)*	1(ref.)*	1(ref.)*	1(ref.)	1(ref.)*
1–4 years	1.00 (0.97–1.03)	1.02 (0.99–1.05)	**1.08 (1.03–1.14)**	**1.04 (1.01–1.06)**	1.02 (0.99–1.05)	**1.07 (1.02–1.12)**
5+years	0.95 (0.91–0.99)	**0.89 (0.84–0.94)**	0.95 (0.87–1.05)	**1.05 (1.02–1.09)**	0.96 (0.92–0.99)	1.02 (0.96–1.09)
Log likelihood	298,766.59	220,057.93	86,445.375	299,365.34	231,992.69	91,251.2
Pseudo R² Nagelkerke	0.297	0.316	0.326	0.352	0.381	0.385

Note: Only the effects of the commuting variable are displayed in this table, but all the variables shown in Table 6.2 were included in the models and the relationship between these variables and separation had the same direction in both models

*Indicates that the whole variable is significant at $p \leq 0.01$, that is, it shows whether or not the model as a whole becomes better when the variable is included in it

[a]Odds ratio, values > 1 show increased odds. Significant values ($p \leq 0.01$) are marked with bold text

[b]95% confidence interval

rienced as great and the utility as low, and could thus affect the relationship more negatively than would be the case if, for example, commuting was chosen to receive a higher wage.

CONCLUDING DISCUSSION

Commuting is salient in many people's everyday life. Although it has its advantages, numerous disadvantages can make the long-distance commuting lifestyle difficult. As the numbers of long-distance commuters are growing, it stands to reason that more couples will face the pros and cons of commuting in their daily routines. While some couples have to handle the consequences of long-distance commuting temporarily, for only a number of years, others will face and even adapt to a more long-term commuting lifestyle.

There are several reasons why long-distance commuting might be expected to affect separation rates, in either direction, and it appears that the separation effect is ambiguous. First, it seems as if the first years of long-distance commuting may be the most challenging for a relationship. Secondly, for those couples for whom long-distance commuting has been part of their lives for more than a few years, separation rates are lower. This offers support for the customization process, whereby couples with time manage to adjust their lifestyle and develop strategies to handle the many presumable costs (both social and economic) of a long commute. It may be the case that for many the mobility choice of long-distance commuting is strategic and more socially sustainable than other alternatives, such as migrating and losing social networks and/or is a result of migrating and keeping one's old job and so on, and they, therefore, have a higher tolerance for the commuting stress. On the other hand, a selection effect can explain that those separating relatively soon after choosing this mobility strategy lack these abilities to customize to the daily costs or would have separated anyway.

When other factors in the event history analysis are controlled for, the results also reflect gender differences. Male long-distance commuters can expect to separate to a higher extent than non-commuting men when commuting less than five years. However, in general, much fewer women long-distance commute at all, reflecting that the gender expectations and structural constraints about breadwinning and parenthood still prevails. These norms are likely to make it more problematic for women than for men to long-distance commute, both practically and emotionally, which can cause women to consider long-distance commuting as a threat to their relationship. Nevertheless, those women long-distance commuting for many years run a lower risk of seeing their relationship broken. It may be the case that these women manage better than men do to adjust to the commuting lifestyle. A possible explanation for this may be that these women are not in a relationship with traditional views on household roles, that is, that the woman should shoulder the main part of the domestic work and childcare and be the second wage earner. They may, then, in contrast to the traditional gender differences in commuting, place more weight on labour market aspirations and less on domestic responsibilities, have support in household-related duties at home and therefore tolerate longer commuting times. Further, this suggests a more modern lifestyle in which the traditional gender differences in long-distance commuting are abandoned.

The results also showed that the long-distance commuter effect on relationships varied depending not only on whether or not the long-distance commuter was a woman but also on what geographical context the couple lived in. An explanation for these geographical differences can be that there are different causes for starting long-distance commuting depending on where you live, which has different outcomes on relationships.

To summarize, one might expect the social costs of long-distance commuting to reduce the quality of a relationship in many ways and thus increase the risk of separation. The statistical results from these analyses not only confirm such assumptions about social costs but can also reveal other and more unexpected results regarding the effects of long-distance commuting on relationships. Depending on whether it is the man or the woman who long-distance commutes, couples seem to handle the social costs of long-distance commuting in different ways. When it is the woman who is the persistent long-distance commuter, couples seem to manage to create a sustainable work-life balance although one partner is a long-distance commuter.

The causality between long-distance commuting and separation was not possible to establish within this study, however. How the long-distance commuter and his/her spouse can tolerate or manage the costs of long-distance commuting is of course influenced by external circumstances and societal changes such as illness, unemployment and changes in household composition. Furthermore, in order to understand why some couples can adapt to a commuting lifestyle and others separate after a number of years, there is a need for future research applying other methods such as interviews and questionnaire surveys. How the social cost of long-distance commuting is experienced by both the commuter himself or herself and the spouse is of importance for planners and policymakers striving to promote socially sustainable transportation patterns.

NOTES

1. The average travel time for a journey (work, business, school or leisure trip) in Sweden is 42 minutes for 27 kilometres according to the Swedish National Travel Survey (Swedish Institute for Transport and Communications Analysis 2007). Therefore, 30 Euclidean kilometres is assumed to take at least 45 minutes on average.
2. As cohabiting is a very common form of relationship in Sweden, compared to many other western countries, no distinction is made between married and cohabiting couples. About one-third of all couples in Sweden are cohabitating (Statistics Sweden 2003) and out of all new couples between 1999 and 2001 67% were cohabiting (with common children) and 33% were married (Statistics Sweden 2011).
3. Only spouses with common children are registered as cohabiting in the data.
4. The number of individuals and separations are not equal among women and men because some individuals are married to/cohabiting with or separated from either non-residential Swedes or non-Swedish persons. These individuals are not included in the analyses.
5. Only couples with known duration of marriage/cohabitation are accounted for here, n=412,730.

REFERENCES

Allison, P. D. (1982). Discrete-time methods for the analysis of event histories. *Sociological Methodology, 13*, 61–98.

Alonso, W. (1964). *Location and land use: Toward a general theory of land rent.* Cambridge, MA: Harvard University, I, 6.

Andersson, G. (1997). The impact of children on divorce risks of Swedish women. *European Journal of Population, 13*, 109–145.

Barker, L., & Connolly, D. (2006). Long distance commuting in Scotland, Scottish Household Survey/Transport Research Planning Group Topic Report, Scottish Executive Social Research.

Brueckner, J. K. (2000). Urban sprawl: Diagnosis and remedies. *International Regional Science Review, 23*(2), 160–171.

Cassidy, T. (1992). Commuting-related stress: Consequences and implications. *Employee Counseling Today, 4*(2), 15–21.

Costa, S. (2004, May 11). *A review of long-distance commuting: Implications for northern mining communities.* Paper presented at the Canadian Institute of Mining, Metallurgy and Petroleum and AGM Conference, Edmonton.

Costa, G., Pickup, L., & Di-Martino, V. (1988). Commuting – A further stress factor for working people; evidence from the European Community, I. A. Review. *International Archives of Occupational and Environmental Health, 60*(5), 371–376.

Evans, G. W., Wener, R. E., & Phillips, D. (2002). The morning rush hour: Predictability and commuter stress. *Environment and Behaviour, 34,* 521–530.

Fisher, P. A., & Malmberg, G. (2001). Settle people don't move: On life course and (im)mobility in Sweden. *International Journal of Population Geography, 7,* 357–371.

Flood, M., & Barbato, C. (2005). Off to work. Commuting in Australia. Discussion paper Number 78, Australian Institute.

Friberg, T., Brusman, M., & Nilsson, M. (2004). Persontransporternas "vita fläckar". Om arbetspendling med kollektivtrafik ur ett jämställdhetsperspektiv. Centrum för kommunstrategiska studier, Linköpings Universitet.

Fults, K. K. (2010). *A time perspective on gendered travel differences in Sweden.* Licentiate Thesis in Infrastructure, Department of Transport and Economics, Royal Institute of Technology, Stockholm, Sweden.

Gatersleben, B., & Uzzell, D. (2007). Affective appraisals of the daily commute. Comparing perceptions of drivers, cyclists, walkers, and users of public transport. *Environment and Behavior, 39*(3), 416–431.

Gottholmseder, G., Nowotny, K., Pruckner, G. J., & Theurl, E. (2009). Stress perception and commuting. *Health Economics, 18,* 559–576.

Green, A. E., Hogarth, T., & Shackleton, R. E. (1999). Longer distance commuting as a substitute for migration in Britain: A review of trends issues and implications. *International Journal of Population Geography, 5,* 49–67.

Hoem, J. M. (1997). *The impact of the first child on family stability.* Stockholm Research Reports in Demography No. 119, Stockholm University. http://www.suda.su.se/SRRD/srrd119.doc

Houghton, D. S. (1993). Long-distance commuting: A new approach to mining in Australia. *The Geographical Journal, 159*(3), 281–290.

Jain, J., & Lyons, G. (2008). The gift of travel time. *Journal of Transport Geography, 16*(2), 81–89.

Kluger, A. N. (1998). Commute variability and strain. *Journal of Organizational Behaviour, 19*(2), 147–165.

Koslowsky, M., Aizer, A., & Krausz, M. (1996). Stressor and personal variables in the commuting experience. *International Journal of Manpower, 17*(3), 4–14.

Krieger, M., & Fernandez, E. (2006). Too much or too little long-distance mobility in Europe? EU policies to promote and restrict mobility. Technical report, European Foundation for the Improvement of Living and Working Conditions. http://www.eurofound.eu.int/docs/areas/populationandsociety/mobility-4paper2006.pdf

Lidström, A. (2006). Commuting and citizen participation in Swedish city-regions. *Political Studies, 54,* 865–888.

Lyons, G., & Urry, J. (2005). Travel time use in the information age. *Transportation Research Part A: Policy and Practice, 39*(2–3), 257–276.

Lyons, G., & Chatterjee, K. (2008). A human perspective on the daily commute: Costs, benefits and trade-offs. *Transport Reviews, 28*(2), 181–198.

Mokhtarian, P., & Salomon, I. (2001a). How derived is the demand of travel? Some conceptual and measurement considerations. *Transportation Research Part A, 35*, 695–719.

Mokhtarian, P., & Salomon, I. (2001b). Understanding the demand for travel: It's not purely "derived". *Innovations, 14*(4), 355–380.

Mokhtarian, P. L., Salomon, I., & Redmond, L. S. (2001). Understanding the demand for travel: It's not purely 'derived'. *Innovation: The European Journal of Social Science Research, 14*(4), 355–380.

Muth, R. F. (1969). *Cities and housing; the spatial pattern of urban residential land use*. Chicago: University of Chicago Press.

Ory, D., & Mokhtarian, P. (2005). When is getting there half the fun? Modeling the liking for travel. *Transportation Research Part A, 39*(2–3), 97–123.

Páez, A., & Whalen, K. (2010). Enjoyment of commute: A comparison of different transportation modes. *Transportation Research Part A, 44*, 537–549.

Plaut, P. (2006). The intra-household choices regarding commuting and housing. *Transportation Research Part A, 40*, 561–571.

Pocock, B. (2003). *The work/life collision*. Sydney: The Federation Press.

Putnam, R. D. (2001). *Bowling alone: The collapse and revival of American community*. New York: Simon and Schuster.

Redmond, L., & Mokhtarian, P. (2001). The positive utility of the commute: Modeling ideal commuting time and relative desired commute amount. *Transportation, 28*, 179–205.

Reneland, M. (1998). Befolkningens avstånd till service. GIS-projektet Tillgänglighet i svenska städer 1980 och 1995. Rapport 1998:5, Göteborg, Sverige: STACTH Stads- och trafikplanering Arkitektur Chalmers Tekniska Högskola.

Renkow, M., & Hoover, D. (2000). Commuting, migration, and rural-urban population dynamics. *Journal of Regional Science, 40*(2), 261–287.

Roberts, J., Hudson, R., & Dolan, P. (2009). It's driving her mad: gender differences in the effects of commuting on psychological well-being, Working papers 2009009. The University of Sheffield, Department of Economics, revised May 2009.

Rotter, J. C., Barnett, D. E., & Fawcett, M. L. (1998). On the road again: Dual-career commuter relationships. *The Family Journal: Counselling and Therapy for Couples and Families, 6*(1), 46–48.

Rüger, H., & Ruppenthal, S. (2010). Advantages and disadvantages of job-related spatial mobility. In N. Schneider & B. Collet (Eds.), *Mobile living across Europe II. Causes and consequences of job-related spatial mobility in cross- national perspective* (pp. 69–93). Opladen/Farmington Hills: Barbara Budrich Publishers.

Russo, G., Reggiani, A., & Nijkamp, P. (2007). Spatial activity and labour market patterns: A connectivity analysis of commuting flows in Germany. *The Annals of Regional Science, 41*(4), 789–811.

Sandow, E., & Westin, K. (2010). The persevering commuter – Duration of long-distance commuting. *Transport Research Part A, 44*, 433–445.

Singer, J. D., & Willett, J. B. (1993). It's about time: Using discrete-time survival analysis to study duration and the timing of events. *Journal of Educational and Behavioral Statistics, 18*(2), 155–195.

So, K. S., Orazem, P. F., & Otto, D. M. (2001). The effects of housing prices, wages and commuting time on joint residential and job location choices. *American Journal of Agricultural Economics, 83*(4), 1036–1048.

Statistics Sweden. (2003). Single and cohabiting adults in the population register, tax register and in reality, background statistics to population- and welfare statistics 2003:11, Statistics Sweden.

Statistics Sweden. (2011). Kan yrket förklara skilsmässan? *SCB tidskrift Välfärd, 2*, 11–14.

Storey, K. (2001). Fly-in/fly-out and fly-over: Mining and regional development in Western Australia. *Australian Geographer, 32*(2), 133–148.

Stutzer, A., & Frey, B. S. (2008). Commuting and life satisfaction in Germany. *Scandinavian Journal of Economics, 110*(2), 339–366.

Swedish Agency for Economic and Regional Growth. (2010). Classification of Sweden's regions. http://www.tillvaxtverket.se/huvudmeny/faktaochstatistik/omregionalutveckling/regionfamiljer.se. 2010-09-04.

Swedish Institute for Transport and Communications Analysis. (2007). The national travel survey RES 2005/06. Swedish Institute for Transport and Communications Analysis (SIKA), No. 2007:19.

Tolley, R. (1996). Green campuses: Cutting the environmental costs of commuting. *Journal of Transport Geography, 4*(3), 213–217.

Travisi, C. M., Camagni, R., & Nijkamp, P. (2010). Impacts of urban sprawl and commuting: A modeling study of Italy. *Journal of Transport Geography, 18*(3), 382–392.

van der Klis, M., & Karsten, L. (2009). The commuter family as a geographical adaptive strategy for the work-family balance. *Community, Work & Family, 12*(3), 339–354.

van der Klis, M., & Mulder, C. H. (2008). Beyond the trailing spouse: The commuter partnership as an alternative to family migration. *Journal of Housing and the Built Environment, 23*(1), 1–19.

van Ommeren, J. (1996). *Commuting and relocation of jobs and residences: A search perspective*. PhD thesis, Virje Universiteit Amsterdam, Amsterdam.

Van Ommeren, J., Van den Berg, G. J., & Gorter, C. (2000). Estimating the marginal willingness to pay for commuting. *Journal of Regional Science, 40*(3), 541–563.

Vilhelmson, B. (2002). Rörlighet och förankring. Geografiska aspekter på människors välfärd. Göteborg: Göteborgs universitet: Kulturgeografiska institutionen Handelshögskolan. Chorus 2002:1.

Wachs, M., Taylor, B. D., Levine, N., & Ong, P. (1993). The changing commute: A case study of the jobs-housing relationship over time. *Urban Studies, 30*(10), 1711–1730.

Zax, J. S. (1991). The substitution between moves and quits. *The Economic Journal, 101*(409), 1510–1521.

CHAPTER 7

Measuring *Mobilities of Care*, a Challenge for Transport Agendas

Inés Sánchez de Madariaga and Elena Zucchini

IMPLICATIONS OF CARE ACTIVITIES AND UNPAID WORK FOR TRANSPORT RESEARCH

Since the late 1970s, a number of studies have analysed the different transportation patterns of men and women, particularly in the USA and the UK, usually providing quantitative empirical analyses of travel patterns. Various works by Rosenbloom (1993, 1995, 1996, 1998) provide early overviews of relevant issues. A few of these authors have focused on the relationship between gender, time poverty and transport (Turner and Grieco 1998; Pickup 1988). Others have looked at issues such as low-income women and transport (Blumenberg 2016; Hamilton 1999; Lucas 2012), public transport and the needs of women (Guiliano 1979), household structure and mobility patterns (Bernard et al. 1997), gender differences in travel to work (Wekerle & Rutherford 1987; O'Brien & Shemilt 2003; Blumen 1994; Hanson and Pratt 1995), gender gaps in commuting (Rosenbloom & Burns 1993;

I. S. de Madariaga (✉)
Universidad Politécnica de Madrid, Madrid, Spain
e-mail: i.smadariaga@upm.es

E. Zucchini
Universidad Politécnica de Madrid, Madrid, Spain

© The Author(s) 2019   145
C. L. Scholten, T. Joelsson (eds.), *Integrating Gender into Transport Planning*, https://doi.org/10.1007/978-3-030-05042-9_7

Gordon et al. 1989; Crane 2007) and gender differences in travel lifecycles (Collins & Tisdell 2002). Stiewe (2012), Schultz and Gilbert (1996) and Wekerle (1992) provided early insights into safety issues in transport. Polk (1996) and Root et al. (2000) looked at the relationship between gender, transport and environmental sustainability, while Oxley and Charlton (2011), among others, analysed gender differences in driving cessation, and Grieco and McQuaid (2012) considered the transaction costs and policy gaps in transport when taking gender into consideration. Universal design (Audirac 2008) is an important aspect for ensuring women's needs and gender dimensions are properly addressed when designing transport systems. Other more recent contributions include the work of Turner et al. for the European Parliament of 2006, the first special issue on gender in the city published by the Town Planning Review (Sánchez de Madariaga & Neuman 2016) and that of Loukaitou (2016), looking at both the first and third worlds. Uteng and Cresswell (2008) provide wider outlooks on gendered mobilities beyond the urban including accross countries.

A few contributions have also examined institutional policies, gender equality in the transport sector and good practice (Rohmer 2007; Swedish Road Administration 2009, 2010; Transport for London 2007; Gender Equality Unit 2004; Hamilton et al. 2005; Mayor of London 2004; Greed et al. 2003; DoT 2000; Bofill et al. 1988). An important source of research on this topic has been the proceedings from the *Women in Transport* conferences organized since the late 1970s by the US Federal Highway Administration (2009) and the Transportation Research Board of the US National Academies (FHA TRB 1996, 2004, 2009, 2011). A particularly relevant key contribution to this now rich body of specialized literature has been the description of the *chained trips* typically performed by persons, mostly women, with young children (Pickup 1985; Grieco et al. 1989; Hanson 1980a, b; Rosenbloom 1989; Blumen 1994; McGuckin & Murakami 2005; McGuckin & Nakamoto 2005).

While this significant body of research has provided important insights into the different travel patterns of men and women, resulting mostly from their gender roles, it is also true that this greater understanding of gender differences in travel has not had much impact on how transport systems are built and operated. It is also true that this body of research on women and gender in transport has not—as yet—had a significant impact upon mainstream transport research, specialized programmes on the topic in higher education, transport planning or policy. It generally remains a marginal topic in the transport literature and in the education of future transport engineers and planners.

However, a proper understanding of gendered travel patterns is a key issue for how people use transport systems, which need to become a core part of educational programmes and the professional literature. As the focus of transport policy begins to put a greater emphasis on people, accessibility and service, somewhat moving away from earlier approaches that were more single-mindedly centred on infrastructure provision, the focus on users' accessibility will bring a greater acknowledgement of the gender dimensions of transport use. As providing services to all users and accessibility for all become core approaches, it will be necessary to look at gender as a key dimension to ensure equality of access to transport as a basic element for ensuring equality of access for all, irrespective of gender, to urban goods and services.

THE *MOBILITY OF CARE*

Earlier work by one co-author of this chapter provides a new concept, the *mobility of care* (Sánchez de Madariaga 2009, 2010, 2013a, b), which has the potential to provide the practical and operational underpinnings needed to sustain a new approach to data gathering, and hence policy definition and implementation, for a better integration of gender considerations into transport planning and management.

The concept *mobility of care* acknowledges the need to evaluate and identify daily travel related to care activities. Care is understood as the mostly unpaid work carried out by adults having responsibility for children and other non-physically autonomous individuals, as well as those activities needed for the upkeep of the home. Official data provided by public institutions in Europe and the US (Eurostat 2016; US Bureau of Labour Statistics 2011) show significant gender differences with women consistently spending more time on care-related activities across countries. This gender gap does not decrease significantly with time, although there are significant differences among countries.

The *mobility of care* includes trips made for the following purposes related to daily life: escorting children to school, to sports and to other extra-curricular activities; doing non-leisure shopping; errands in public offices; visiting and escorting sick and elderly relatives, and so on. Such mostly unpaid activities imply travelling to specific locations in the city, at specific times during the day, using the available transport systems, under certain conditions of price, ergonomics and safety. Importantly, they need to be combined and made compatible with work in paid employment.

The overall travel implied by all of these activities can be quite difficult, lengthy and cumbersome. As the above-mentioned literature has shown, lack of appropriate means of transport can impose significant limitations on people's lives, including forcing individuals to work part time, being stripped of leisure and personal time, or having to completely renounce working in paid employment.

Surveys performed by public bodies in charge of transportation policy and investment do not normally allow for a proper measuring of care-related travel (Sánchez de Madariaga 2009, 2010, 2013a, b). An analysis by this author of the main transport surveys conducted in Spain demonstrates several sources of gender bias in the way in which they gather, interpret, analyse and visually represent data.[1] Several of the Spanish surveys analysed in this study do not take into consideration short trips on foot of less than 1 km or 15 min; chained trips are not taken into consideration; trips related to care are hidden under other headings such as 'visiting', 'strolling', 'leisure', 'escorting' or 'other'. Members of the household in one of the surveys were to be defined as either breadwinners or housewives, with no additional possibilities for self-definition. Graphical representations included a number of visual distortions that resulted in the overrepresentation of the weight of employment-related travel.

All these sources of bias and omissions add up to the redistribution of care-related trips under numerous small headings, several of which are more akin to the concept of personal leisure than to the notion of care as defined above: the unpaid work performed by adults for the care of others and the upkeep of the home. This redistribution under many small categories, together with the significant number of trips not counted, leads to the underquantification, undervaluation and invisibility of travel related to care.

Assuming that the *mobility of care*, understood as a new umbrella category encompassing the various kinds of trip needed for the reproduction of life, represents an important group of general mobility, earlier work by Sánchez de Madariaga (2013a) roughly estimated that such trips might account for a volume, measured in the number of trips, close to that of trips linked to employment. This rough estimation was made on the basis of statistical data provided by existing transport surveys in Spain, but not on the actual measuring of such trips with specifically designed methodologies.

The following graph represents the results of this rough estimation of the number of trips made for purposes of care in metropolitan areas in Spain from existing data and under the following assumptions (Sánchez de Madariaga 2013a). The data used correspond to that of an important and extensive nationwide survey undertaken by the Spanish National Ministry of Infrastructure in 2006, called Movilia, that covers all metropolitan areas in the country (Ministerio de Fomento 2006–2007). A rather arbitrary assumption was made regarding how many of the trips under the existing headings of this survey could be considered as care trips, rather than leisure or personal trips: two-thirds of shopping trips, one-third of strolling trips, all escorting trips, one-third of visits and one-third of other. The data were re-coded under these hypothetical assumptions to offer an estimate of the size of care-related mobility. The results can be seen in Fig. 7.1. In this hypothetical model, care as a purpose for travel becomes second only to employment. If leisure is also understood as an umbrella concept including strolling and visits, in addition to travel to locations

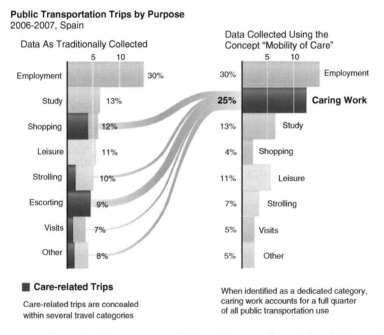

Fig. 7.1 A hypothetical estimation of the percentage of care-related trips, made by re-coding existing data from an official survey. (Source: Sánchez de Madariaga 2013a, b)

technically designated as leisure, then care and leisure, as wide categories for travel, appear to be of similar size to each other and very close to the size of employment travel.

Bringing the notion of care to the forefront of transport research and policy will allow us to challenge existing gender biases in both the construction of knowledge and the policy decisions that derive from it.

Implicit gender assumptions underpin planning concepts; for example, the key idea of spaces of *work* being separate from the spaces of *living* (Sánchez de Madariaga 2004). Such concepts are built upon the typically male personal experience of the individual who has no care responsibilities and whose main experience of the city is that of resting in the home and working for a salary in paid employment. It does not reflect the personal experiences of those, mostly women, who take care of others, most of whom also work in paid employment. When looked at from a gender perspective, we see that residential areas are not really spaces of rest, but places where a lot of work (unpaid) is undertaken on a daily basis (Ullmann 2013).

As we deconstruct the gender-implicit assumptions underlying transport and planning as academic, professional and institutional fields of endeavour, we need to look more closely at women's travel needs and the factors that affect them. Women's mobility is influenced by a number of issues. The first is linked to land-use planning and accessibility, meaning the degree of difficulty in reaching a place in terms of distance and time spent for the trip. Accessibility is defined by three factors: (a) the location of the milestones that define the route to be taken; (b) the infrastructure existing between them; (c) the means of transport available for the journey (Ilarráz 2006).

Safety is another key element in understanding women's mobility (Wekerle 1992). Many women feel threatened in badly kept, dark and dirty places and on public transport where staff or customers exhibit anti-social or aggressive behaviour. Women's mobility is particularly conditioned by their gender roles, which imply a double workload and, consequently, chained trips to fulfil the many tasks of daily life. Additionally, many of these trips tend to be non-predictable.

Time and space cannot be considered as two separate entities when taking these chains of tasks into account. The opportunity to carry out a specific chain of activities depends not only on the location and distance between the various tasks, but also on the possibility of traversing the distance between them, or 'links of the chain' (Gepken 2002) within a specific period of time. The daily chains of tasks are influenced by many variables, including location of the home vis-à-vis employment and services, type of employment (full or part time), income level, marital status, number and age of children, existing means of transport and whether there is responsibility for elderly, disabled or sick relatives.

Defining an Appropriate Methodology for the Analysis of the *Mobility of Care*

The *mobility of care* recognizes the need to evaluate and make visible the daily travel resulting from care work. It provides a framework for considering the relevant variables that affect daily life and consequently the way in which people use the city and move within it.

As explained above, this section proposes a methodology specifically designed to test the hypothetical assumptions made in previous work regarding the extent of the *mobility of care* (Schiebinger et al. 2013).[2] When designing the survey, we paid particular attention to the definition of the categories of data to be collected and on the wording of questions to be asked, so as to avoid the gender biases and implicit assumptions that often prevent the identification of mobility patterns related to care tasks (Sánchez de Madariaga 2009).

The first criterion applied was to create an analytical umbrella category for the *mobility of care*. A second criterion involved the counting of all trips, regardless of the means of transport, the duration or the purpose of each trip. A third criterion was to provide a wide and detailed enumeration of the specific activities that qualify as care, so as to clearly identify the trips made for that purpose, and to properly separate them from the trips made for other purposes, and particularly from leisure, leisure shopping, strolling or visiting. It was also important to study the main socio-economic variables that influence travel and to make the relevant crossovers between them and the variables of time, means of transport and purpose.

In accordance with this, the survey included four main analytical categories: the first is employment, the second is care—subdivided into many smaller and specific categories relating to the care of others or the upkeep of the home, the third is study, and the last is leisure—also understood as an umbrella concept that includes travel to what planning officially calls leisure spaces, and also trips made for personal reasons (Fig. 7.2).

This methodology was applied in a survey designed and carried out in the metropolitan area of Madrid. The variables considered were as follows:

General Survey - Mobility of Care		
Sex:		F M
Age:		

	Single	
	In a relationship	
Who are you?	Married	
	Divorced	
	Widow	
	Refused	
	Yes	
Do you have children?	Not	
	Refused	
	Alone	
	Family	
	children (how many?)	
Who do you live with?	*Husband/wife*	
(please tick all that apply to	*Elderly relatives (how many?)*	
you)	*Parents (how many?)*	
	Other relatives	
	Partner	
	Friends	
	Refused	
	Urban	
	Post-code	
where do you live?	Suburban	
	Rural/Exurban	
	Refused	
	Housing or rental price	
	The local schools	
	Location to a job site	
What was the most	Location a school site	
important reason you		
chose your current home	Location to shopping, entertainment, restaurants	
location?	Location to a social, religious, civic, cultural or recreational facility	
(please tick the 3 most	Transit access	
important: 1st; 2nd; 3rd)	Closeness to relatives or friends	
	Other (Please Specify)	
	Refused	

Fig. 7.2 General survey: *mobility of care.* (Source: I. Sánchez de Madariaga, E. Zucchini)

- Employment status
- Family responsibilities
- Marital status
- Access to a private vehicle
- Security and risk perception
- Disability
- Economic resources
- Level of education
- Ethnicity
- Place of residence
- Reasons for the selection of residential location
- Existing means of transport
- Prices and pricing policy

In order to counteract gender biases and stereotypes, conscious or unconscious, it was important to define previous instructions addressing both the interviewers and the interviewees.[3] These instructions explained the purpose of the study, and what we mean, for example, by the term *care* and the activities it encompasses, as well as the nuances that distinguish *employment*, as formal, paid work in the economy, from a wider understanding of activities which could be considered *work* from a gender perspective, which also includes unpaid home-related tasks, and are different from leisure and personal purposes. The instructions were also necessary for the interviewer to maintain a firm and non-invasive position with respect to the interviewees, so as to avoid unwanted situations that might at best contaminate results and at worst jeopardize the outcome of the interview (Converse 1970).

In surveying women on gender-related issues, the selection of people to carry out the interview is relevant. In particularly sensitive contexts, the culture and education of the interviewees could inhibit the discussion of certain topics with men. On the other hand, people who are in charge of the survey have to be trained on gender issues, so that they become aware of their role, in order to reduce both potential gender biases and the influencing of answers (Mayntz et al. 1976).

The instructions for the interviewers, who had to explain to the interviewees the purpose of the survey and how to achieve this, were the following:

With this survey we want to assess and make visible the daily travel associated with care work. We understand care work as the unpaid labour performed by adults for or with children and other dependants, including labour related to the upkeep of a household.

Once the interviewees have filled out the "General Survey", they must also complete the second part of the survey, which involves specific answers about the trips they make during the day.

You must ask about every trip they make during the day, to recognize the care trips.

Instructions:

Ask about the trips they made yesterday. The starting place is HOME, and the day finishes at 11.00 pm.

Fill out one column for EACH location they go to and the PURPOSE; if necessary, ask them to explain what kind of activity they do in this location. If uncertain as to whether to include a location at which they stop, include it.

Record ALL locations visited.

Record ALL the means of transport (including walking).

Record the EXACT time at which they arrive at and leave each location.

Record the activities (what they did) at each location.

If they park their car and walk for MORE than five minutes to their destination, record the type of transport as car first and then walk. If they walk for MORE than five minutes from a bus to their destination, record their transport as bus first and then walk.

The detailed purposes related to care work included under the umbrella category are the following:

Childcare

- *Escorting to:*

    1. *School*
    2. *Nursery*
    3. *Other activity (social:* e.g. *Playground, …)*
    4. *More frequently used services (Doctor, Pharmacy, …)*
    5. *Other Services (Library, Shopping, …)*
    6. *Employment*
    7. *Strolling*

- *Activity for children:*

    1. *School*
    2. *Shopping*

3. *Visit*
4. *Turn Around*
5. *Primary Services (Hospital, Pharmacy, ...)*
6. *Other activities (preparing meals, homework...)*

Attention to other dependent individuals (e.g. parents, relatives or someone else not a family member)

- *Escorting to:*

  1. *Shopping*
  2. *Other activity (Social, Religious...)*
  3. *Services (Hospital, Pharmacy, ...)*
  4. *Strolling*
  5. *Errand*

- *Activity performed for other individuals:*

  1. *Visit*
  2. *Shopping*
  3. *Pick up/Drop off*
  4. *Errand*
  5. *More frequently used services (Doctor, Pharmacy, ...)*
  6. *Other activities (preparing meals, house cleaning, ...)*

Unpaid Home-Related Activities

We designed a *face-to-face* questionnaire that paid particular attention to avoiding gender bias and to using neutral language without ambiguous terms or technicalities. We differentiated the trips associated with paid work from the trips associated with care work, and we used questions with multiple possible answers, to collect all the possible options of response according to the different situations of women and men. The purpose of each trip was asked about in sufficient detail that trips related to care activities could be clearly distinguished from trips resulting from paid work, leisure, family and friends' visits, personal trips and study. The survey includes trips and chain trips carried out while escorting others, and all types of trips, irrespective of duration or distance, in order to avoid missing out short trips on foot (Fig. 7.3).

Questionnaire - Mobility of care				
Sex:				
Age:				
TRIP Nº ..	TRIP Nº ..	TRIP Nº ..	TRIP Nº ..	TRIP Nº ..
From Home to Place .. ...:... am/pm	From Place .... to ...:... am/pm	From Place .... to ...:... am/pm	From Place .... to ...:... am/pm	From Place .... to ...:... am/pm
Location	Location	Location	Location	Location
Home	Home	Home	Home	Home
Work	Work	Work	Work	Work
Other...	Other...	Other...	Other...	Other...
Purpose (tick all that apply to you)	Purpose (tick all that apply to you)	Purpose (tick all that apply to you)	Purpose (tick all that apply to you)	Purpose (tick all that apply to you)
Attend Childcare	Attend Childcare	Attend Childcare	Attend Childcare	Attend Childcare
* *Your Children*	* *Your Children*	* *Your Children*	* *Your Children*	* *Your Children*
* *Children of your family*	* *Children of your family*	* *Children of your family*	* *Children of your family*	* *Children of your family*
* *Children out of your family*	* *Children out of your family*	* *Children out of your family*	* *Children out of your family*	* *Children out of your family*
Attend Someone else	Attend Someone else	Attend Someone else	Attend Someone else	Attend Someone else
* *Parents*	* *Parents*	* *Parents*	* *Parents*	* *Parents*
* *Someone of your family*	* *Someone of your family*	* *Someone of your family*	* *Someone of your family*	* *Someone of your family*
* *Someone out of your family*	* *Someone out of your family*	* *Someone out of your family*	* *Someone out of your family*	* *Someone out of your family*
Work	Work	Work	Work	Work
Escorting	Escorting	Escorting	Escorting	Escorting
Personal Business	Personal Business	Personal Business	Personal Business	Personal Business
Shopping	Shopping	Shopping	Shopping	Shopping
Strolling	Strolling	Strolling	Strolling	Strolling
Visit	Visit	Visit	Visit	Visit
Health (Pharmacy, Doctor, Hospital...)	Health (Pharmacy, Doctor, Hospital...)	Health (Pharmacy, Doctor, Hospital...)	Health (Pharmacy, Doctor, Hospital...)	Health (Pharmacy, Doctor, Hospital...)
Errand	Errand	Errand	Errand	Errand
Recreation	Recreation	Recreation	Recreation	Recreation
Social	Social	Social	Social	Social
Other....	Other....	Other....	Other....	Other....
Transport	Transport	Transport	Transport	Transport
Car	Car	Car	Car	Car
Motorcycle	Motorcycle	Motorcycle	Motorcycle	Motorcycle
Bycicle	Bycicle	Bycicle	Bycicle	Bycicle
Walk	Walk	Walk	Walk	Walk
Bus	Bus	Bus	Bus	Bus
Metro	Metro	Metro	Metro	Metro
Taxi	Taxi	Taxi	Taxi	Taxi
Train	Train	Train	Train	Train
Other....	Other....	Other....	Other....	Other....
Distance	Distance	Distance	Distance	Distance
0-0,99 km	0-0,99 km	0-0,99 km	0-0,99 km	0-0,99 km
1-1,99 Km	1-1,99 Km	1-1,99 Km	1-1,99 Km	1-1,99 Km
2-2,99 Km	2-2,99 Km	2-2,99 Km	2-2,99 Km	2-2,99 Km
3-4,99 Km	3-4,99 Km	3-4,99 Km	3-4,99 Km	3-4,99 Km
5-9,99 km	5-9,99 km	5-9,99 km	5-9,99 km	5-9,99 km
10-19,99 Km	10-19,99 Km	10-19,99 Km	10-19,99 Km	10-19,99 Km
more than 20 Km	more than 20 Km	more than 20 Km	more than 20 Km	more than 20 Km
Time	Time	Time	Time	Time
0-10 min	0-10 min	0-10 min	0-10 min	0-10 min
11-20 min	11-20 min	11-20 min	11-20 min	11-20 min
21-30 min	21-30 min	21-30 min	21-30 min	21-30 min
31-45 min	31-45 min	31-45 min	31-45 min	31-45 min
46-60 min	46-60 min	46-60 min	46-60 min	46-60 min
more than 60 min	more than 60 min	more than 60 min	more than 60 min	more than 60 min

Fig. 7.3 Face-to-face questionnaire—*mobility of care*. (Source: I. Sánchez de Madariaga, E. Zucchini)

The selection of the sample is a key aspect. To obtain quality data, it is important to achieve a sample that will be representative of pivotal variables at the basis of inequalities between men and women. In this regard, the sample has to be large enough to gather the necessary information to be stratified by sex.

We used a quota sample, which is frequently applied in marketing or public opinion surveys. Through this type of sample, we specify the defining characteristics of the people we want to study. In this case, we fixed age and sex. In accordance with the proportionality of the population of Madrid, we delimited a quota for each characteristic of the sample. Thus, as Madrid's population is composed of 51% women and 49% men, we utilized a sample made up of 50% women and 50% men. The number of inhabitants is 3,165,000 people, and admitting an error of 3.5% and a confidence level of 95%, we interviewed a sample of 800 people, aged between 30 and 45 years. The respondents were selected within 100 census tracts of Madrid, and eight people were interviewed for each section. Through the survey, we analysed 3323 trips made by the interviewed population within a period of 24 hours.

The sections were chosen in five districts of the metropolitan area of Madrid, two of which are located in the centre, one within the boundaries of the most central area of the city and the other two in the suburbs of the metropolitan area. It was decided to study these specific districts in order to better understand how location influences people's ways of moving. Additionally, the location of residences is a good indicator of income, itself an important variable influencing the selection of means of transport.

QUANTITATIVE ANALYSIS OF THE *MOBILITY OF CARE* IN THE CITY OF MADRID

The analysis of the data collected through this survey offers a number of very significant results. The first and most important result is the overall quantification of the trips made for the purpose of caring for others and the upkeep of the home. Figure 7.4 shows on the right-hand side the percentage of trips made for care activities for the population aged 30–45 years, as compared on the left with the traditional categories normally used in transport surveys. Figures 7.5 and 7.6 further show the data disaggregated by sex.

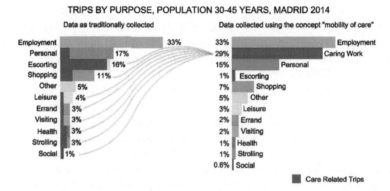

Fig. 7.4 Trips by purpose, population 30–45 years, Madrid 2014. Traditional categories and umbrella category *mobility of care*. (Source: I. Sánchez de Madariaga, E. Zucchini)

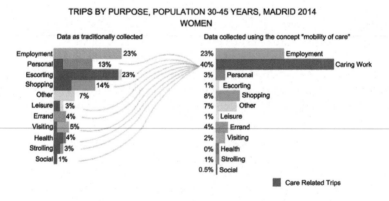

Fig. 7.5 Trips made by women by purpose, population 30–45 years, Madrid 2014. Traditional categories and umbrella category *mobility of care*. (Source: I. Sánchez de Madariaga, E. Zucchini)

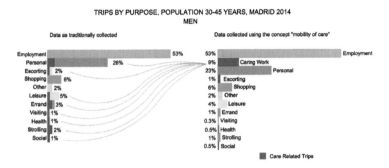

Fig. 7.6 Trips made by men by purpose, population 30–45 years, Madrid 2014. Traditional categories and umbrella category *mobility of care*. (Source: I. Sánchez de Madariaga, E. Zucchini)

The first conclusion is the validation of our hypothesis; that is, the total number of trips made for the purposes of care is close to the total number of trips made for reasons of employment, that is, 29% as compared to 33% for this age segment, a four-point difference that is practically equal to the five-point difference in the estimation made in our hypothetical model (Sánchez de Madariaga 2013a, b).

A second conclusion is the startling differences by sex, with women undertaking the lion's share of care-related trips, 40% of their trips compared to 9% of men's. Men do most of the employment-related travel, making up 53% of their trips, 30 points above women's, at 23%. Men also make many more trips for leisure and personal issues, with relatively high percentages at 23% and 4% respectively. Women significantly make a very small number of trips for leisure, 3%, and personal purposes, 1%. It is worth noting that for women, trips recorded as personal by conventional means of categorization become trips done to attend and take care of others, when questions are asked in this kind of detail, and the category is introduced into the surveys.

A third conclusion is the confirmation of the relevance of using a new umbrella category that makes visible the size of care-related mobility, which is currently hidden within a variety of common categories normally used by transport surveys around the world. A graphical representation of the same set of data using the old categories to the left, and the new categories to the right, helps to illustrate this point. The importance of good-quality, non-biased, noise-free graphical visualizations of quantitative data is highly relevant here. Equally important is the introduction of gender-aware notions that create new conceptualizations of the realities we are attempting to understand in ways that more accurately reflect the lived experiences of women and men.

The 800 people surveyed made a total of 3323 trips, of which 66% were made by women, with an average exceeding 5 trips a day, and the remaining 34% were made by men, with an average of 2.8 trips a day (Fig. 7.7).

The very significant differences between men and women are obviously related to the age segment of the sample, 30–45 years. This is the segment of the population in which certain gender-relevant activities are especially pronounced: marriage, jump-starting and consolidating professional careers, motherhood and caring for young children. As explained above, narrowing down the age segment in the sample was necessary for practical reasons of feasibility of the research on which this chapter is based.

Further research to identify variations by age that looks at younger and older individuals is needed. Different age segments of the population will most probably show different patterns.

Figure 7.8 shows the different means of transport used by men and women out of the total trips made by each group. Women make 45% of their trips by car, 24% on foot, 20% by collective transport (shared almost equally by bus and metro) and only 5% by motorcycle. Men use mostly cars (30%), the metro (23%) and motorcycles (24%).

There are many reasons to validate these data. As stated above, one is the entrance of women into the labour market, which, together with the increase in standards of living of the population, has allowed women of the middle and upper classes to have easier access to a second car. Because of

Fig. 7.7 Number of trips, percentage distribution by sex. Population 30–45 years, Madrid 2014 (total trips 3323). (Source: I. Sánchez de Madariaga, E. Zucchini)

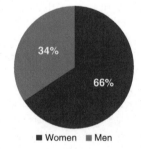

■ Women ■ Men

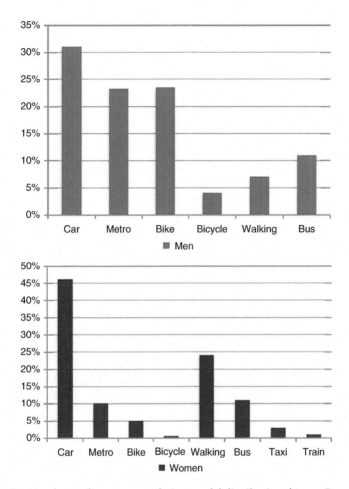

Fig. 7.8 Number and percentage of trips, modal distribution, by sex. Population 30–45 years, Madrid 2014. (Source: I. Sánchez de Madariaga, E. Zucchini. Note: No. of trips men 1145; no. of trips women 2178; total no. of trips 3323)

the structure of contemporary cities, in which young children cannot travel alone, and where educational, health, sports services and so on are located at non-walkable distances, the car has become the most important and indispensable means of transport for mothers with young children. Many of the trips made by car by women in this segment of the population are trips made escorting children to school and other places.

But the main reasons why the car is better adapted to the needs of care are that it is faster than other means of transport, more flexible and makes it easier, sometimes being the only possible means, to attend all the increasingly diverse and complex activities needed for daily life. Women who have children under 18 make more than 70% (calculated on 66% of total trips) of the trips analysed, and men in the same situation make only 41% (calculated on 34% of total trips).

The fact of having children, and therefore attending to their travel needs, indicates an important difference between men's and women's mobility patterns, basically considering the variable time. For women, trips of less than 10 min represent 54% of the total, and 75% of these trips are made by women who have dependent children. By contrast, for men trips of less than 10 min make up only 21% of the total, and men with dependent children make only 10% of these trips (Figs. 7.9 and 7.10).

The analysis indicates that women's trips turn out to be shorter. Considering the massive use of the car, these data show that women who have access to a car are using it to chain together more and shorter trips.

Important factors affecting mobility are those activities in which people engage on a daily basis. People in employment perform a very high proportion of trips, almost 90% of the total. When we look at the distribution of activities related to care, there is a significant difference regarding the type of employment, whether it is part time or full time. This turns out to be particularly relevant from a gender perspective because of gender differences in access to full employment (Fig. 7.11).

The trips made by people who work part time are equivalent in the number to the trips of those who work full time. Most people working full time are men. Conversely, and this is a very important issue for understanding the gender dimensions of travel, it is women who hold the vast majority of part-time jobs (Fig. 7.12).

Out of the total number of trips made by women, part-time workers make up 61% and the remaining 39% is divided into almost equal parts between those who work full time and those who are unemployed.

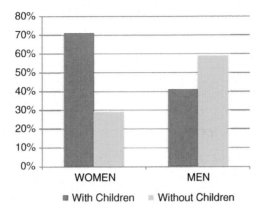

Fig. 7.9 Trips made by men and women, with and without dependent children, as the percentage of the total for each group. Population 30–45 years, Madrid 2014. (Source: I. Sánchez de Madariaga, E. Zucchini)

Fig. 7.10 Trips according to employment. Population 30–45 years, Madrid 2014. (Source: I. Sánchez de Madariaga, E. Zucchini)

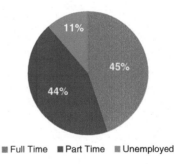

It is very important to underline the low percentage of men working part time, in contrast to the high percentage of women who do. These figures, together with those about trips segregated by type of employment, are an additional confirmation that women choose a part-time job to be able to carry out household and care tasks.

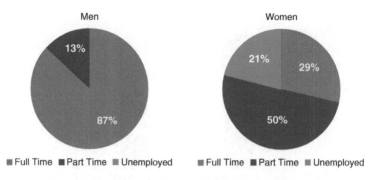

Fig. 7.11   Type of employment, distribution by sex. Population 30–45 years, Madrid 2014. (Source: I. Sánchez de Madariaga, E. Zucchini)

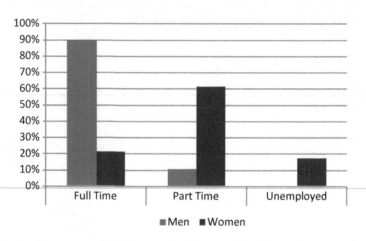

Fig. 7.12   Percentage of trips by type of employment, distribution by sex. Population 30–45 years, Madrid 2014. (Source: I. Sánchez de Madariaga, E. Zucchini)

As women continue to maintain their role as caregivers, the proportion of trips related to activities performed for social reproduction among women in paid work is three times higher than among men in the same situation.

To understand the different patterns of mobility, it is also significant to look at who performs the more specific types of activities included under the umbrella category of care. These specific activities directly influence the different modal split for women and men. Different distribution of these tasks among men and women explains why men in this segment of the population use more means of transport, since their duties are not specifically concentrated in one activity, but their dedication is shared between the different categories (Fig. 7.5). Instead, women need more flexible transport to be able to implement their tasks (Fig. 7.6).

Among the different categories within care tasks, for both men and women, the highest percentage of trips corresponds to escorting, which is an activity strictly linked to the use of a car. Escorting for men represents 1.35% of their total number of trips (1145) and for women 20% of their total number of trips (2178).

The category of shopping, specifically non-leisure shopping, is another activity which in most contemporary urban settings has to be carried out using a car, because of new consumption models and the structure of commercial space. Daily shopping in the local neighbourhood has been replaced by bigger and less frequent trips to shopping centres. In our data, shopping appears to be women's work representing 5.50% of their trips and 0% of men's.

Regarding the time spent travelling and the distance travelled, men make longer trips than women, but at the end of the day, women spend more time moving around. Although their trips are shorter, they make many more of them, and they are chained more often (Fig. 7.13).

In addition, the distribution of the average duration of trips according to purpose and gender is influenced by the different locations of the homes of individuals. Both men and women living in the urban centre generally travel by public transport and go longer distances. In contrast, among those living in more peripheral areas, there are important gender differences: women spend less time travelling than men, because women are forced to use private transport due to the reduced availability of public transport.

Time is a key variable in the analysis of the mobility of care. More than 40% of care trips have a duration of less than 10 min, because many of these trips are made by people for whom the neighbourhood becomes the place of *work* (Fig. 7.14).

Regarding differences in the intensity of mobility and the changes in modal split according to purpose and location, we found the following. In

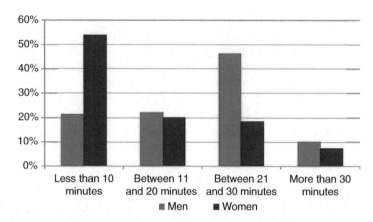

Fig. 7.13   Average duration of trips, distribution by sex. Population 30–45 years, Madrid 2014. (Source: I. Sánchez de Madariaga, E. Zucchini)

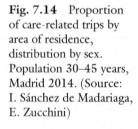

Fig. 7.14   Proportion of care-related trips by area of residence, distribution by sex. Population 30–45 years, Madrid 2014. (Source: I. Sánchez de Madariaga, E. Zucchini)

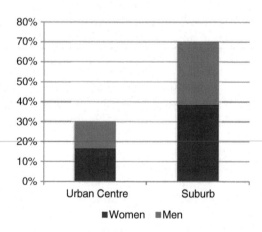

the central districts of Madrid, the discrepancy is reduced and dedication to care work does not register a big difference according to gender. In the central area, women do 55% of care-related trips and men 45% (just a 10% gap); in contrast, in the suburbs, women do 71% of care trips and men do only 29% (a 42% gap).

It is also interesting to compare the number of trips made for care purposes according to the type of residential neighbourhood. The total number of trips for care activities almost doubles for people living in suburban areas, as compared to those living in the centre. It is also significant to see that in these areas, the gender gap multiplies several times: while in the centre the gap is only 10%, in suburban areas the gap increases fourfold to over 40%. In these peripheral areas, women carry out the majority of care tasks.

These data also clarify why women are the main users of cars in the overall sample. It is the care tasks of women living in non-central areas that explain the high use of cars for the overall population of Madrid in this age segment. In peripheral areas, it is almost impossible to move around without using private transport, as services, in most cases, are not located within walking distance and in these types of neighbourhood, public transport is not an efficient option to carry out daily life.

Conclusion: Implications for Transport Policy Agendas

A number of useful conclusions and recommendations for transport policy and research can be drawn from this investigation. We would like to point out some that we find particularly interesting.

We propose that the umbrella category *mobility of care* should be used as an analytical category in research, planning and policy. Using this concept as a starting point to thoroughly describe people's mobility will allow the creation of a broader knowledge base about patterns of mobility, behaviour and needs, in which travel associated with the needs of daily living is made visible and properly counted. In addition, this will improve functional concepts and transport policies to make them more equitable and responsive to gender needs, thus benefiting society as a whole.

We recommend that quantitative analyses are designed carefully, looking to avoid potential gender biases that can occur at any stage during the research process, from the early stages of designing a survey, to the implementation stage, to the analysis of results and the graphical representation of results. Quantitative analyses should obviously include all trips, the means of transport, times and purposes that are at the basis of each trip. These purposes have to be defined by taking into account and detailing appropriately the many different activities related to care tasks that are mostly performed by women on a daily basis and are normally not considered in transport surveys.

We acknowledge the significant importance that ethnographic data have for understanding mobility patterns (Baylina 1996). For example, the fact that people place greater value on care tasks than work when deciding where to live is a highly relevant finding. However, because of the overvaluation of anything dealing with the economy in our contemporary societies, and the comparative undervaluation attached to women's activities, urban and transport policies often base their programmes on assumptions and priorities that are primarily derived from the sphere of formal economic activity.

Despite the fact that this research has been developed from a relatively small sample of people of a limited age range, our results show that certain mobility patterns that planners and policy-makers take for granted as universal represent only the mobility of a specific part of the population. In order to better align decision-making with the diversity of people's needs, it is necessary that public policies correctly evaluate the comparative weight of the mobility of care vis-à-vis mobility related to employment.

Obviously, a proper understanding of employment-related mobility is a very important aspect of policy-making, in particular when planning for peak hours. However, this should not obscure the fact that care-related mobility represents an almost equal share of the total number of trips. The specific requirements of this type of mobility need to be taken into consideration when designing, investing in and managing transport systems. An important tool that could make a significant change in the way in which mobility is evaluated is social participation. Participatory processes need to be designed in ways that ensure women's voices are included.

The next steps in this research should involve applying the methodology proposed here to bigger samples across broader segments of the population, which would allow crossing data from other age segments. It would be particularly interesting to carry out this investigation with people older than 60 years, who in many countries are the ones performing a significant share of care tasks, both for minors (grandchildren) and for people aged over 80 years.

Likewise, it would be relevant to develop empirical studies measuring the *mobility of care* in other countries and cities with different socioeconomic make-ups, with different levels of income, in places where safety is one of the central issues conditioning accessibility and in locations with different transport systems. This kind of study and investigation would provide a better understanding of how care activities influence the use of transport systems. They will also demonstrate how the design and management of transport systems influences the care economy.

NOTES

1. This study analysed gender biases in the four main Spanish transportation surveys: Movilia, conducted at the national level by the Ministry of Infrastructure, which includes data from all metropolitan areas (Ministerio de Fomento 2007); two surveys conducted by the regional government of Catalonia (Generalitat de Catalunya 2006), and one conducted by the regional government of Madrid (Consorcio de Transportes de Madrid 2004; Sánchez de Madariaga 2009).
2. The previous research showing the innovative concept *mobility of care* has been showcased as a case study in the EU–US project Gendered Innovations http://genderedinnovations.stanford.edu/case-studies/transportation. html. This project provides analytical methods and case studies on how to introduce gender dimensions into the content of research in technological and medical fields. The design of this methodology and its application to the region of Madrid is a result of the PhD dissertation by Elena Zucchini (2016): "*Género y transporte: análisis de la movilidad del cuidado como punto de partida para construir una base de conocimiento más amplia de los patrones de movilidad. El caso de Madrid*", Universidad Politécnica de Madrid, supervised by Inés Sánchez de Madariaga.
3. On the issue of gender bias in research and how to counteract it, see the Gendered Innovations Project mentioned above.

REFERENCES

Audirac, I. (2008). *Universal design and accessible transit systems: Facts to consider when updating or expanding your transit system*. Washington, DC: Project Action.

Baylina, M. (1996). *Metodología cualitativa y estudios de geografía y género* [Qualitative methodologies and studies in gender geography]. Universidad Autónoma de Barcelona. Departament de Geografia.

Bernard, A., Seguin, A.-M., & Bussiere, Y. (1997). Household structure and mobility patterns of women in O-D surveys: Methods and results based on the case studies of Montreal and Paris. In *Women's travel issues: Proceedings from the second national conference*, October 1996, FHWA, US Department of Transportation, TRB, pp. 249–266.

Blumen, I. (1994). Gender differences in the journey to work. *Urban Geography, 15*(3), 223–245.

Blumenberg, E. (2016). Why low-income women in the US still need automobiles. *Town Planning Review, 87*(5), 525–545.

Bofill Levi, A., Dumenjó Martí, R.-M., & Segura Soriano, I. (1988). *Las Mujeres y la ciudad manual de recomendaciones para una concepción del entorno habitado desde el punto de vista del género*. Barcelona: Fundació Maria Aurelia Capmany.

Collins, D., & Tisdell, C. (2002). Gender and differences in travel life cycles. *Journal of Travel Research, 41*, 133–143.

Consorcio de Transportes de Madrid. (2004). *Encuesta de movilidad de Madrid EDM*. Madrid: Consorcio de Transportes de Madrid.

Converse, P. (1970). Attitude and no attitudes: Continuation of a dialog. In E. Tafte (Ed.), *The quantitative analysis of social problems* (pp. 168–189). Reading: Addison-Wesley.

Crane, R. (2007). Is there a quiet revolution in women's travel? Revisiting the gender gap in commuting. *Journal of the American Planning Association, 73*(3), 298–316.

Department for Transport. (2000). *Women and public transport: The checklist.* London: DETR.

Federal Highway Administration FHA. (1996, 2009, 2011). *Research on women's issues in transportation.* Conference proceedings. Washington, DC: Transport Research Board.

Gender Equality Unit. (2004). *How to incorporate gender equality into infrastructure, housing, transport, urban development, youth services.* Factsheets reports 2000 04. London: NDP.

Generalitat de Catalunya. (2006). *Enquesta de mobilitat quotidiana* (EMQ06) Barcelona: Generalitat de Catalunya.

Gepken, F. (2002). Como incorporar una perspectiva de género en la practica corriente del planeamiento, in Sánchez de MAdariaga I. (Dir) *Segundo Seminario Internacional de Género y Urbanismo*, Madrid: Universidad Politécnica de Madrid.

Gordon, P., Kumar, A., & Richardson, W. (1989). Gender differences in metropolitan travel. *Regional Studies, 23*(6), 499–510.

Greed, C., Devis, L., Brown, C., & Dure, S. (2003). *Gender equality and plan making.* London: Royal Town Planning Institute.

Grieco, M., & McQuaid, R. (2012). Special issue gender and transport: Transaction costs, competing claims and transport policy gaps. *Research in Transportation Economics, 34*, 1–86.

Grieco, M., Pickup, L., & Whipp, R. (1989). *Gender and transport: Employment and the impact of travel constraints.* Aldershot: Avebury.

Guiliano, G. (1979). Public transportation and the travel needs of women. *Traffic Quarterly, 33*(4), 607–616.

Hamilton, K. (1999). Women and transport: Disadvantage and the gender divide. *Town and Country Planning, 68*(10), 318–319.

Hamilton, K., Jenkins, L., Hodgson, F., & Turner, J. (2005). Promoting gender equality in transport. *Working Paper Series*, no. 34. Manchester: Equal Opportunities Commission.

Hanson, S. (1980a). The importance of multi-purpose journey to work in urban travel behavior. *Transportation, 9*, 229–248.

Hanson, S. (1980b). Spatial diversification and multipurpose travel: Implications for choice theory. *Geographical Analysis, 12*(3), 245–257.

Hanson, S., & Pratt, G. (1995). *Gender, work and space.* New York: Routledge.

Eurostat. (2016). Harmonized European Time Use Survey (HETUS). *Prepared tables, main activities (2-digit level) by sex and country.* Eurostat http://www. h2.scb.se/tus/tus/.

Ilárraz, I. (2006). Movilidad sostenible y equidad de género, *Zerbitzuan Journal, 40*, 61–66.

Loukaitou-Sideris, A. (2016). A gendered view of mobility and transport. *Town Planning Review, 87*(5), 547–565.

Lucas, K. (2012). Transport and social exclusion: Where are we now? *Transport Policy, 20*, 105–113.

Mayntz, R., Holm, K., & Hubner, P. (1976). *Introduction to empirical sociology.* London: Penguin.

Mayor of London. (2004). *Expanding horizons: Transport for London's women's action plan:* London: Mayor of London.

McGuckin, N., & Murakami, E. (2005). *Examining trip-chaining behavior: A comparison of men and women.* Washington, DC: United States Department of Transportation.

McGuckin, N., & Nakamoto, Y. (2005). Differences in trip chaining by men and women. In United States National Research Council (Ed.), *Research on women's issues in transportation report of a conference, Vol. II: Technical papers.* Washington, DC: Government Publishing Office (GPO).

Ministerio de Fomento. (2007). *Encuesta de Movilidad de las Personas Residentes. Movilia 2006.* Madrid: Ministerio de Fomento.

O'Brien, M., & Shemilt, I. (2003). *Working fathers: Earning and caring.* Manchester: Equal Opportunities Commission.

Oxley, J., & Charlton, J. (2011). Gender differences in attitudes to and mobility impacts of driving cessation. In S. Herbel & D. Gaines (Eds.), *Women's issues in transportation: Summary of the fourth international conference, 27–30 October 2009, Irvine, CA* (Vol. 2, pp. 64–73). Washington, DC: United States National Research Council (NRC) Transportation Research Board.

Pickup, L. (1985). Women's gender-role and its influence on travel behavior. *Built Environment, 10*, 61–68.

Pickup, L. (1988). Hard to get around: A study of women's travel mobility. In J. Little, L. Peake, & P. Richardson (Eds.), *Women in cities: Gender and the urban environment.* New York: New York University Press.

Polk, M. (1996). *Swedish men and women's mobility patterns: Issues of social equality and ecological sustainability.* US Department of Transportation, Federal Highway. Administration, Washington DC, http://www.fhwa.dot.gov/ohim/womens/chap11.pdf

Rohmer, H. (2007). Gender mainstreaming European transport research and policy: Building the knowledge base and mapping good practices. Available at http://koensforskning.soc.ku.dk/projekter/transgen/

Root, A., Schintler, L., & Button, K. (2000). Women, travel and the idea of sustainable transport. *Transport Reviews, 20*(3), 369–383.

Rosenbloom, S. (1989). Trip chaining behaviour: A comparative and cross cultural analysis of the travel patterns of working mothers. In M. Grieco, L. Pickup, & R. Whipp (Eds.), *Gender, transport and employment.* Aldershot: Avebury.

Rosenbloom, S. (1993). Women's travel patterns at various stages of their lives. In C. Katz & J. Monk (Eds.), *Full circles: Geographies of women over the life course.* London: Routledge.

Rosenbloom, S. (1995). Travel by women. In *Federal Highway Administration, nationwide personal transportation survey,* Demographic Special Reports.

Rosenbloom, S. (1996). Women's travel issues in Federal Highway Administration *Women Travel Issues, Proceedings from the second national conference.* Washington, DC: Transport Research Board.

Rosenbloom, S. (1998). *Trends in women's travel patterns.* Berkeley: The University of California Transportation Center.

Rosenbloom, S., & Burns, E. (1993). *Gender differences in commuter travel in Tucson (Arizona).* Berkeley: University of California Transportation Center.

Sánchez de Madariaga, I. (2004). *Urbanismo con perspectiva de género.* Sevilla: Fondo Social Europeo - Junta de Andalucía.

Sánchez de Madariaga, I. (2009). Transporte metropolitano y grupos sociales: propuestas para una mejor planificación. Report for CEDEX. Madrid: Ministry of Infrastructure.

Sánchez de Madariaga, I. (2010). Housing, mobility and planning for equality in diversity: Cities, gender and dependence. In *VVAA social housing and city* (pp. 177–197). Madrid: Ministerio de Vivienda.

Sánchez de Madariaga, I. (2013a). From women in transport to gender in transport: Challenging conceptual frameworks for improved policy making. *Journal of International Affairs, 67,* 43–66.

Sánchez de Madariaga, I. (2013b). The mobility of care: Introducing new concepts in urban transportation. In I. Sánchez de Madariaga & M. Roberts (Eds.), *Fair shared cities: The impact of gender planning in Europe.* Aldershot: Ashgate.

Sánchez de Madariaga, I., & Neuman, M. (2016). Mainstreaming gender in the city. *Town Planning Review, 87*(5), 493–504.

Schiebinger, L., Klinge, I., Sánchez de Madariaga, I., & Schraudner, M. (Eds.). (2013). *Gender innovations in science, health and medicine, engineering and environment* (launched 2011: genderinnovations.stanford.edu).

Schultz, D., & Gilbert, S. (1996). Women and transit security: A new look at an old issue. *Proceedings of the Women's Travel Issues Second National Conference,* 25–27 October, Baltimore.

Stiewe, M. (2012). Gender and mobility: Everyday mobility during changing gender relations. In *26th AESOP annual congress*, 11–15 July, Ankara: Association of European Schools of Planning.

Swedish Road Administration (SRA). (2009). *The road transport sector: Sectoral report 2008*. Stockholm: SRA.

Swedish Road Administration (SRA). (2010). *The road transport sector: Sectoral report 2009*. Stockholm: SRA.

Transport for London. (2007). *Gender equality scheme*. London: Group Publishing.

Transportation Research Board. (2004, 2009, 2011). *Research on women's issues in transportation conference: Report of conference proceedings, TRB, Washington, DC* Transportation Research Board of the US National Academies, http://www.trb.org/Main/Blurbs/164708.aspx

Turner, J., & Grieco, M. (1998). *Gender and time poverty: The neglected social policy implications of gender for time, transport and travel*. Paper presented at International Conference on Time Use. Luneberg: University of Luneberg. http://www.geocities.com/margaret_grieco/womenont/time/.html

Turner J., Hamilton K., & Spitzner M. (2006). *Women and transport report. Brussels:* European Parliament.

Ullmann, F. (2013). Choreography of life. In I. Sánchez de Madariaga & M. Roberts (Eds.), *Fair shared cities: The impact of gender planning in Europe*. London/New York: Ashgate.

United States Bureau of Labour Statistics. (2011). *American time use survey: 2010 results*. Washington, DC: United States Department of Labour.

United States Federal Highway Administration. (2009). *National household travel survey (NHTS): Summary of travel and trends*. Washington, DC: Government Publishing Office (GPO).

Uteng, T., & Cresswell, T. (Eds.). (2008). *Gendered mobilities*. New York: Ashgate.

Wekerle, G. (1992). *A working guide to planning and designing safer cities*. Toronto: City of Toronto Planning and Development Department.

Wekerle, G., & Rutherford, B. (1987). Employed women in the suburbs: Transportation disadvantage in a car-centered environment. *Alternatives: Perspectives on Society, Technology, and Environment, 14*(3–4), 49–54.

PART IV

Transport Planning Beyond Gender Stereotypes

The 'I' in Sustainable Planning: Constructions of Users Within Municipal Planning for Sustainable Mobility

Malin Henriksson

INTRODUCTION

An important quest for a future feminist transport agenda is that planning policy and practice must reflect a diverse set of experiences. Historically, the planning profession has prioritized the interests and needs of privileged men (Feinstein and Servon 2005). Planning theory, practice and education has typically been dominated by white, middle-class men who have consequently been viewed as experts on the built environment (Sandercock and Forsyth 2005; Snyder 1995). There are many examples of how the lived experiences of women and marginalized groups have been neglected in the search for 'the common good' (Sandercock 1998). Today, we see that the planning profession is not dominated by men in such numbers any more. As in many academic professions, women are even outnumbering men (for the Swedish context, see Friberg 2006; Larsson and Jalakas 2008). Does this mean that the planning practices and ideals of today will be more equal? Feminist scholars argue that even

M. Henriksson (✉)
VTI, Linköping, Sweden
e-mail: malin.henriksson@vti.se

© The Author(s) 2019
C. L. Scholten, T. Joelsson (eds.), *Integrating Gender into Transport Planning*, https://doi.org/10.1007/978-3-030-05042-9_8

though representation matters to some degree, the values, interests and beliefs that are supported in the workplace environment are still crucial for the outcome of planning (Greed 1999; Doan 2011; see also Greed, this collection). Planners, regardless of gender or background, are affected by the surrounding milieu. As Greed (1999) argues, those who easily adopt the normative ideals, whether they are women or men, will have easier access to powerful settings and positions. Thus, planning cultures will shape planners' views on knowledge and their epistemologies, as well as defining what and who are regarded as important or unimportant, interesting or uninteresting and what questions will have high or low priority. To plan is to impose certain realities on space (Greed 1994) and to bring particular aspects of expertise and knowledge into the limelight.

The intention of this chapter is to discuss how normative ideas about the subjects of planning are expressed by professionals and how the findings of this research can help feminist planning practices to develop. I take my point of departure from the ongoing move from planning for motorized transport to planning for more sustainable modes of travel, known as 'the modal shift' (Banister 2008). As a result, changing individual travel behaviour (i.e. 'mobility management' or 'travel demand management') has emerged as an important line of action within municipal planning (Friman et al. 2013; Hrelja et al. 2013; Marsden and Rye 2010). The practice of travel demand management is underpinned by the notion that travel behaviour can be steered in a given direction and calls for knowledge about human behaviour.

Shove (2010) argues that what she calls the 'ABC' model has become a dominant paradigm within climate-change policy. This model is based on psychological literature grounded in theories of planned behaviour. Here, behavioural change is thought to depend upon values and attitudes (the A), which are believed to drive the kinds of behaviour (the B) that individuals choose (the C) to adopt (Shove 2010: 1274). The ABC paradigm dominates transport behaviour research, as well as transport policies aimed at influencing travel patterns (Davies et al. 2014). According to Heisserer (2013), there are several reasons for the popularity of this 'actor-centric' approach. One important factor is the aura of 'common sense' that surrounds these theories, which attracts a diverse audience from different (educational) backgrounds. Moreover, they can easily be adopted in quantitative inquiries that produce 'hard' numerical data. The area of transport policy is dominated by actors with a background in science and/or engineering, where numerical data is regarded as 'real' knowledge and qualitative accounts do not fit into the knowledge production process (ibid., see also Grange 2010; Hrelja and Antonsson 2012). This generates important questions: do the diverse set of actors involved in planning for sustainable mobility have access to knowledge about travel behaviour and how to change it? How do they interpret this knowledge?

As pointed out by, among others, Shove et al. (2012), travel behaviour is not merely a reflection of the attitudes and values of users of the transport system. Reinforcing this understanding dismantles the political aspects of public policy in general and transport planning in particular, making individuals responsible for the much-needed transition to a sustainable society (Levy 2013; Paterson and Stipple 2010; Cupples and Ridley 2008). However, travel behaviour is embedded in social, political and cultural contexts. Primarily, the mass production of affordable cars during the twentieth century combined with the redesigning of cities towards car-friendly (and bicycle-hostile) infrastructure has created a system dominated by car-based mobility (Böhm et al. 2006; Paterson 2008; Koglin 2013). An example of how power structures both cause and reflect behaviour is the gendered aspects of mobility (Uteng and Creswell 2008). For instance, research on gender and mobility reveals that women's daily trips are often more complex than men's due to household responsibilities (see Sánchez de Madariaga & Zucchini's contribution in this collection). When their needs are not supported by public transport, users must expend a considerable amount of emotional and physical effort to overcome distances between workplaces and living spaces (Hjorthol 2000; Friberg 1993). Furthermore, there are gendered differences in experiences of safety and fear of violence (Koskela and Pain 2000; Valentine 1989).

The embedded nature of travel behaviour makes it relevant to study how ideas about such behaviour are shaped in institutional settings. When working with transport planning and mobility management, professionals are explicitly aiming at influencing travel behaviour. This gives rise to several important questions that I intend to comment on in this chapter. *Who* is asked to change their travel behaviour? When professionals talk about changing behaviour, are there any special groups of travellers that attract their attention? What kind of mobility practices are included in their understandings, and what practices are excluded?

In the following, I will argue that how professionals understand sustainable mobility is closely interlinked with their own mobility practices. My contribution to the subject will be a discussion of how professionals in a Swedish municipality understand sustainable mobility in relation to their own experience as professionals *and* as users of transport systems. In line with Bacchi (1999), framings of this particular policy area will shape the future orientation of sustainable mobility policy and will include some citizens' experiences of mobility while excluding others. The overall aim of this chapter is to explore how municipal officials understand users of the transport system in relation to sustainable mobility. More specifically, I will investigate how professionals construct user identities and how the outcome of my analysis can be of value in transport planning.

The chapter is organized as follows: I will first introduce the setting of the study, followed by my two theoretical concepts, 'the I-methodology' and 'user representations'. In the empirical sections, I will discuss how professionals construct particular user identities in relation to their own notions of sustainable mobility. I will highlight two user representations: the motorist and the cyclist. The final analysis includes questions about what kinds of travel practices professionals take for granted, establish as normal and regard as legitimate to talk about.

BACKGROUND

The empirical material is based on a study that was conducted in 2009–2013 in Helsingborg, the eighth largest municipality in Sweden with approximately 130,000 inhabitants, of whom 90,000 live within the city.[1] As in many other Swedish cities, environmental standards for emissions from traffic have been exceeded for several years. Among other issues, Helsingborg is characterized by a relatively small city centre dominated by several large traffic routes, which leads to heavy traffic. To face these challenges, the municipality has adopted several objectives to reduce car use. These include plans for increased cycling and public transport as well as projects aiming to change travel behaviour. Within the municipal organization, the urban planning office and the environmental office share responsibility for implementing and achieving environmental goals. I interviewed 16 professionals from these offices, 6 women and 10 men (for a full list of informants, see appendix). Some were born in the late 1940s, and others in the early 1980s, but most of them were between 30 and 45 years. Their professional backgrounds and education varied. Some identified as planners, while others had a background in traffic engineering or the behavioural sciences. This means that the informants have different professional 'belongings' or 'cultures' and have access to and produce different kinds of knowledge. However, they have in common that they are all, in different ways, responsible for achieving sustainable mobility within the municipality. The diversity of the informants reflects how planning theorist Clara Greed (1999) describes planning practices today, where the activity of planning is something in which a diverse group of actors engage. The challenges facing modern municipalities (such as climate change, de-growth) demand professionals with different sets of skills and backgrounds, she argues (see also Czarniawska 2002).

The interviews evolved around the following themes: the informants' professional backgrounds, how sustainable mobility is incorporated into their work tasks, how they understand the concept of sustainable mobility and lastly how the informants understand the users of the transport system. In the interviews, pictures of different travellers (using cars, bicycles or public transport) were used as stimuli. The purpose of using these stimuli was to accentuate the informants' interpretations of sustainable mobility through inspiring and/or provoking images (Törrönen 2002). The interviews were analysed thematically, with special attention given to how the professionals spoke about users, and what kind of knowledge they employed to make sense of users.

The Construction of Users: A Theoretical Point of Departure

One crucial aspect of city management is to coordinate transport technology (buses, cars, bicycles, etc.) with the infrastructural preconditions of a city. I therefore argue that city managers are dealing with the design and production of technology. Scholars from the field of STS (science and technology studies) highlight that the process of shaping and defining technology is affected by social, political and cultural notions. Gender, ethnicity and disability are examples of factors that limit people's ability to construct and use technology (Star 1991). In relation to the *users of technology*, developments in the field have gone from studying how the technology is used to also including analyses of how the designers and producers of technology configure users themselves (Oudshoorn and Pinch 2008). In line with this reasoning, I argue that two concepts are of particular interest in relation to planning for sustainable mobility: user representations and the I-methodology.

The concept of user representations includes the opinions, motives and desires that certain actors (e.g. planners) ascribe to other actors (e.g. users of transport systems). The process of *user configuration* is a crucial part of technological development (Oudshoorn et al. 2004; Woolgar 1991). To place the emphasis on *semiotic users* rather than real-life users is one way to understand why technology comes to suit certain users but not all. Oudshoorn et al. (2004) are critical towards the early versions of user-centred STS due to the lack of a gender perspective. Oudshoorn and colleagues argue that embedded in user representations are assumptions about the gendered identities of users. In studies on technology, links between masculinity and technology have proved to be strong. Thus, men are often thought of as users and many workplaces where technological development is an important feature are influenced by a masculine culture (ibid.: 40–44) Consequently, some users (often female) are not thought of as users at all. How feminist STS scholars describe male-dominated workplaces bears a strong resemblance to how the planning profession is described within feminist planning theory, which makes this theoretical frame appealing.

An important feature of user configuration is that the actors involved in defining user identities are unaware of the assumptions upon which specific user representations are built. Even when the intention of the designers is to attract 'everyone', this often turns out to be a certain group with specific interests and ideals. These interests often coincide with the interests of the developers of the specific product or infrastructure. This can be explained using the notion of the I-methodology. Oudshoorn et al. define the I-methodology as 'a practice in which designers consider themselves as representations of the users' (2004: 41).

The concept of the I-methodology, coined by French sociologist Madeleine Akrich (1992), includes two understandings. Firstly, Akrich points out that when companies are testing new products, they often specify potential consumer groups from whom their target group could be drawn, such as 'the gadget geek' or 'the businessman on the move'. They use their own ideas or perceptions about these groups to pin down their interests and develop communicative strategies to reach them. Secondly, designers draw upon their own experiences of a product when they test it and adjust the design in line with these experiences. This is particularly apparent when market research or methods of customer analysis are scarce. These phenomena have also been noted in relation to traffic planning (Hrelja and Antonsson 2012). Akrich describes this process as 'when experts take off their professional hats and instead put on lay hats' (1992: 173). The main point of the I-methodology is that designers view themselves as representative of users. Analytically, the I-methodology concept is applied through listening to what kind of knowledge claims professionals refer to when they talk about users, and design for users.

Now, I will explore how the professionals understand sustainable mobility and whom they include in the sustainable mobility discourse. First, I will highlight the kind of travel arrangements that dominate this discourse and then move on to consider what kind of user representations this entails.

UNDERSTANDING SUSTAINABLE MOBILITY: UNDERSTANDING EVERYDAY LIFE

In the interviews, a lot of attention is given to car use. In almost every interview, the professionals agree that from a sustainable mobility perspective, there are too many cars in Helsingborg. It is noticeable that the professionals describe decreased car use as one of the major challenges in the city. Many informants find that a shift from planning for the car to planning against the car is an exciting, new and challenging perspective that they welcome. Henrik, a traffic planner, expresses this view like this:

> You can ask yourself, what city do we want? How many cars do we want to have in the city? How much can we take in order not to feel ill of exhaust fumes and so on? It is happening, I can feel that it's happening a lot right now. There are different sides if you put it like that. Should we do like we've always done, or should we find other options? It's exciting, very exciting.

Henrik describes the car as a problematic feature of urban milieus. His enthusiasm is palpable, but so are the tensions within the municipality. How the future will play out will be determined by power struggles ('there are different sides'), but it is apparent that the hegemony of the car is under question. At the same time, most of the interviewees also acknowledge their own and others' private cars trips. When speaking about what sustainable mobility is, Helena, who works strategically with environmental issues, does not want to describe herself as a sustainable traveller because of how she uses the car in her everyday life. She, like several of the professionals, describes her car use as a necessary evil.

> Sure, on occasion, when my kids were small, I took the pram to the grocery shop. But it took over an hour both ways and you don't do that too often. [...] I feel that a lot of us drive cars because it's so practical. And I drive a car when I need to. But I cycle to school, or to work, or the nursery.

Here, Helena uses her personal experiences in order to understand why people use cars in their everyday lives. Helena represents herself as a layperson rather than an expert with the professional task of reducing car use. Rather than explaining how travelling in the city is affected or can be affected by traffic planning, Helena makes room for her everyday life practices. 'Practical' and timesaving are key words in Helena's account. This is an example of how the I-methodology is used by professionals in order to explain car use.

In the material, trips that are not interpreted as eco-friendly are explained as acceptable from a mundane perspective, where the car is regarded as essential in order to sustain everyday life. The car trips are mainly represented as practical. According to Hampus, most car users have a guilty conscience:

> You must believe that most of us want to do good, I believe so. Regarding attitudes, there are several studies on the topic, well attitudes that reflect that it's a problem to drive a car. Everyone thinks so. Very few people don't admit that their car use is a problem. But at the same time, if someone continues to drive a car it should count as something. That it's difficult in one's everyday life.

This quote is another example of how everyday life is used as an explanation for continued care use. In these quotes from Hampus and Helena, it is obvious that they believe car use is problematic from a professional point of view. But they also share an understanding about the essential role that cars play in people's everyday lives. I understand this as an expression of how their professional identities conflict with their personal experiences of everyday life. The I-methodology thus functions as a linkage between the professional goal to reduce car use and the informants' accounts of users' inability to make more sustainable trips. As a professional, speaking about private experiences could be interpreted as a means to overcome this conflict. The boundary between the professional and the private disappears when sustainable mobility is negotiated. Where the I-methodology is used, boundaries between users and experts are dissolved.

Woolgar (1991) points out that an important feature of constructing user representations is for designers to separate themselves from laymen. To talk about users ('them') as different from designers ('us') is to enhance designers' professional knowledge—'we' know, but 'they' do not. When the professionals negotiate sustainable mobility, this kind of boundary work is not present. Rather than positioning themselves as experts, they position themselves as users.

In the next section, I will present some more examples of how the professionals understand sustainable mobility through the notion of everyday life. Now, I will add specific user representations to the analysis.

'I, A PARENT WHO DRIVES'

Herman is responsible for implementing the political goal of increasing cycling in Helsingborg. When I ask him if this goal also includes changing how the bicycle is perceived, he explains the kind of users that in his opinion have the opportunity to cycle. Herman tells me that even if they (the municipality) want to represent the bicycle as 'a mode of travel with endless possibilities because it's fast and flexible', he also states that you should admit that not everyone has the same opportunity to cycle.

> There are different conditions and, well, we have different jobs. We live in different places and so on. Some of us may even have to drive children and must use the car for work or something like that. There are those awful jobs, you know [laughs]. For the average person, no, but for the ones who are working in the city and it's around five kilometres around here, I think many of them have the possibility to cycle.

What I find intriguing in Herman's reasoning is that he constructs a 'we' who must take the car to work or drive their children to nursery, but also a 'them' who have the opportunity to travel in a sustainable way by bicycle. Herman distinguishes urban residents from residents who live on the outskirts of town, in the suburbs or the countryside. Urban residents are given a bigger responsibility to use the bicycle compared to those who live 'in other places'. They are instead constructed as forced to use the car. Herman could be said to construct spaces for certain people to continue travelling in an unsustainable mode.

Helena's line of reasoning is similar to Herman's arguments. She also links the practices of families with planning practices for sustainable mobility. When we are discussing whether there are any special target groups who should be able to make more eco-friendly trips, Helena states that it is more accurate to speak of 'non-target groups':

> There are groups who drive a lot. Parents with small children drive a lot, to give an example. On the other hand, maybe they haven't got the prerequisite to change the way they travel, in comparison to people who don't have children. Maybe you shouldn't stare yourself blind at the people who drive a lot, because maybe there is no potential for change there.

Here, Helena uses the family as an example of a non-target group. She interprets families with children as a group with no potential for change. Their trips are constructed as stable. Hillevi argues in a similar manner when I ask her to describe a group that needs to change how they travel:

> We've worked a lot with the trips that are shorter than five kilometres, these are the trips we should change. The shorter trips are important to take care of [...] but I also think that if you travel longer distances and if you're a family, maybe it's not that bad that they use the car.

When the professionals imagine different sustainable mobility scenarios, it is often from the point of view of the family. I argue that this is a product of the I-methodology, due to the many examples in which they position themselves as members of families. For example, to drive the kids back and forth to school is a problem with which the professionals have experience. Heidi says:

> And there is also the free school choice,[2] which is great at an individual level, I too have chosen another school, but from an environmental point of view it might not be the best. Because it means that you go back and forth, and you place your children wherever you like. And if the school that you want is not nearby, well then to drive them might be the only option.

This is another example of the ambivalence between the professional goal of reducing car use and personal experience as a driving parent. Heidi presents a specific practice as a problem (driving the kids to school every day) and frames the problem as having negative environmental effects. Even so, Heidi expresses sympathy towards the parents who, according to her, lack other alternatives. She is able to adopt this standpoint by making use of the I-methodology, 'I have chosen another school myself'.

In the interviews, the experiences that the officials share are remarkably similar. They are not only experiences of an everyday life where the car is needed, they are also stories about a certain kind of everyday life. These trips are framed as 'family trips'. Both male and female professionals speak on behalf of families. The (car-dependent) family acts as a role model when professionals construct the car user. This user representation springs from the experiences of the professionals. In line with the I-methodology, the professionals will support, make visible and understand the needs of the family.

'I, A PARENT WHO BIKES'

Not all professionals describe themselves as parents who drive cars. To use the bicycle in everyday life is described in some cases as a viable (and more sustainable) alternative to the car. When these informants describe their bicycle trips, they use different words, filled with different meaning, compared to when they describe car trips. The experience of riding a bike is expressed using positive qualities. It is often male professionals who share examples of how they engage in bicycle trips. Their experiences of cycling make these trips seem almost heroic, and they put themselves forward as 'capable'. For example, when talking about the pictures I brought to the interview, Herman and I discuss the kind of messages about sustainable mobility that it is possible to communicate with citizens. In one of the interviews, Herman asks me whether I want to know more about his personal experiences of sustainable mobility:

Herman: Me, I rode my bike to work with my child every day.
Malin: So, you recognize yourself in that picture[3]?
Herman: I don't know, but I just want to make it clear how I live. As a person, how I relate to it all. So, I cycled 15 kilometres a day.

Herman clearly understands his cycling trips as positive. By telling me about giving his child a ride on his bicycle every day, Herman is enabled to tell me 'how he lives' and how he as a person 'relates to it all'. Herman's tone is challenging when he makes it clear to me that he thinks it is an accomplishment that he used to cycle 15 kilometres a day with his child. This implies that a cycling parent is something special, an approach to parenthood that is out of the ordinary. It also reflects a general tendency in the interviews, namely to talk about cycling as something 'hearty'. However, from the perspective of these professionals, cycling with children is not thought of as something that it is realistic to demand from people in general. I therefore find how Herman speaks about his own experiences as a cyclist to be quite intriguing. Henry shares a similar story:

Well, I don't hesitate to cycle when it's snowy but I cycle a lot and I also use the car a lot and I travel by train rather often. [...] We had, we have two bicycle trailers. So, we used them in order to cycle with both the kids. We went to the nursery that way.

Here, Henry is describing how 'we' (his partner and himself) gave their kids rides to nursery with the help of bicycle trailers. The fact that the childcare trips these men have undertaken by bicycle are constructed as something extra is above all an expression of how accepted car trips are. Car trips are regarded as easy and practical in comparison to bicycle trips. Still, the practice of using the I-methodology is the same as when the car-driving parents and professionals construct sustainable mobility. Furthermore, when Henry and Herman are describing their bicycle practices as heroic, they are adhering to the individualized environmental discourse. Within the frame of this discourse, the individual is constructed as 'good' or 'capable' when she/he uses modes of transport that are considered to be environmentally friendly, such as the bicycle.

In the interviews, it is rare for the professionals to direct criticism towards citizens who travel by car. The car-critical discourse only seems to include planning practices. Thus, when Hampus shares a positive experience from his private life, and relates his practices to the practices of a motorist, he stands out. In the interview with Hampus, he states that the car is embedded in positive cultural understandings and that this makes it even more important to charge sustainable mobility with positive values. As an example, Hampus tells me about the bicycle trips he does with his son in order to promote sustainable mobility:

Yes, well, the clear majority are still learning from their parents. You do what you normally do and if you never used a helmet when you cycle, well then your kids won't use helmets. If you drive them back and forth to different out-of-school activities because you think it saves you some time and because they want you to, then you're not thinking about the whole picture. When they reach a certain age, then they will do the same as you do. If they're not proud of a small carbon footprint it will be impossible to highlight its importance. My son, he is really proud of having cycled for 30 kilometres. If we make an excursion, and we use the GPS on the bicycles, I will ask him how far he thinks we've cycled. Well, more than 20 kilometres he says. No, it's 32 kilometres. Wow, that's a new record!

Hampus' illustration of the successful excursion with his son is characterized by a certain playfulness that cannot be compared with how car trips are described in my material. This could be related to the fact that the bicycle is interpreted as a sustainable means of transport. Therefore, it seems more appropriate to describe the bicycle in a positive manner. To spend time with his son and to show him how much fun it is to engage in sustainable travel can be read as a way to teach him how to be an eco-friendly citizen and taps into the individualized environmental discourse.

Representing What Users?

The two user representations discussed in this chapter are similar in that they both represent different aspects of the everyday mobility of families. Furthermore, they represent different modes of transport. The bicycle is represented as fun, cultivating and healthy, but not always a suitable alternative for parents. Car trips are regarded as boring, but also as problematic from an environmental perspective. Still, the car is understood as a necessity in the everyday lives of families. These conclusions highlight the fact that the topic of planning for sustainable mobility represents an arena where the everyday mobility practices of families are explained and legitimized.

Discussions of the family as a unit often entail traditional gender roles being ascribed to men and women, thus reinforcing gendered stereotypes. From this perspective, it is encouraging to note that the professionals do not present care trips as a female-coded task.[4] Here, the user representation 'parent' is not synonymous with 'mother'. In a longitudinal study about engineers' experiences of engineering, Holth and Mellström (2011) note that it is now more common for male engineers to refer to their own household responsibilities than previously. However, to be a parent in the sustainable mobility setting does not necessarily mean to be caring and therefore to make care trips. It can also be used as a means to emphasize one's own capability, as a father as well as an eco-friendly individual, as in the example of Hampus. Family life can therefore be regarded as an arena within which professionals negotiate both sustainable mobility and gendered identities.

Apart from broadening the notion of parenting from a gender perspective, there are other aspects of the family-oriented norm that mirror gendered power relations. When expressing norms, as the professionals in this study do, values are also attributed to those outside the norm (who could be read as excluded users), and these are projected in parallel. This process is called 'othering' and highlights the importance of listening to 'silences' and 'absences' in relation to the material at hand. Discursive silences make it clear which ideologies a certain discourse is supporting (Holmberg and

Ideland 2009). In this study, silences occur in relation to explanations of mobility and everyday life which are not articulated by the professionals.

The family-oriented user representations normalize having children and being part of a (nuclear?) family. They also normalize the idea that the travel needs of children are most easily fulfilled with the help of cars. However, users who do not have families, or have families that do not include children, are absent. Furthermore, the professionals do not speak on behalf of users who do not have access to a car or are not able to drive a car. Is it possible that there are other reasons to make care trips with a bicycle besides striving for an eco-friendly or healthy lifestyle? This is another discursive silence. The professionals could talk about other kinds of trips than care trips, or other kinds of users. Bradley (2009) links what she calls a Swedish environmental discourse, which, among other things, shapes planners' understanding of what (and who) should be included in an eco-friendly lifestyle or behaviour, with middle-class ideals. According to Bradley, this linkage is visualized when scrutinizing which lifestyles remain unquestioned, and which lifestyles are understood as problematic and in need of change. For example, to recycle and spend time in nature, or to travel by bicycle and public transport, is a part of this discourse. However, other aspects of the lifestyles of many privileged Swedes (such as living in large spaces, travelling by air or car a lot, and shopping as a hobby) are not mentioned in relation to the assumed eco-friendliness. How these professionals normalize a motorized lifestyle could be read as an example of how middle-class ideals are influencing their understandings of sustainable mobility.

When the professionals' framings of sustainable mobility subjects are visualized, as above, one important conclusion is that socioeconomic reasons are not part of a sustainable mobility discourse. Money is not a resource that the semiotic users are missing; rather than money, it is time that is framed as a scarce commodity. Sustainable mobility is regarded as something negotiable, while lifestyles (e.g. driving your children to school) are regarded as stable and without the possibility to change. When sustainable mobility is understood as a self-realizing activity, and a way to promote oneself as an eco-friendly citizen, the freedom to choose the mode of travel is taken for granted. There is a risk that citizens who have little opportunity to make such a choice, who might not have the financial resources to own or use a car, or even to use public transport, are made invisible within this discursive framing of sustainable mobility. This is an example of what is described in the introduction to this book as a process of making the power dimensions of travelling and mobility invisible. The middle class and their way of life are viewed as the norm for society at large. To frame sustainable mobility as a project for the family-oriented middle class is problematic for several reasons. The family is constructed as a starting point for planning when experiences of the family are used in order to renegotiate car trips as part of a non-sustainable mobility. Thus, the nuclear family is excluded from the demand to travel more sustainably. In the final section, I will relate these conclusions to feminist debates within planning theory and make recommendations for how to inform transport planning from a feminist standpoint.

CONCLUDING REMARKS

In this chapter, I have analysed interviews with professionals who are responsible for achieving sustainable mobility goals. One aspect of the interviews that stood out was that the professionals often spoke about their personal travel behaviour, and in several cases also mentioned their private living arrangements in order to explain travel behaviour. Since the interviews largely concerned their professional points of view, this made me wonder whether there was something in the specific context that drove the informants towards their personal experiences. Their remarks could be read as expressions of the individual discourse on travel behaviour, within which it is central that the individual is given the right prerequisites (most often information) to make the right (sustainable) travel choices. In sum, the concept of sustainable mobility is here constructed as something that concerns individual travel choices rather than practices that reflect a car-friendly infrastructure.

This discursive framing does not, however, explain why they talk about their own private experiences instead of inhabitants' experiences of travelling within the municipality, or of theories or ideas on how to change or explain individual travel choices. Akrich (1995) argues that the concept of the I-methodology, a procedure whereby professionals use their own experiences to interpret the needs and desires of users, often comes into play when the professionals lack knowledge, methods or resources to fully understand their target groups. Since, with few exceptions, the professionals in this study did not refer to other sources of knowledge apart from for their own personal experiences when they talked about users, it is reasonable to assume that they lack professional knowledge about travel behaviour and how to steer it, or that they do not regard such knowledge as a valid subject to talk about in this particular setting. The engineering culture, with its strong emphasis on technical aspects rather than the human factor, that traditionally surrounds transport planning environments, does not leave room to explore travel behaviour and how it is embedded in local contexts (see Levy in this volume). Thus, the I-methodology works as a 'lever', a tool which is used to understand something that is outside the professional scope.

So, how can these findings help in developing a feminist practice within transport planning? I started off this chapter by acknowledging the fact that there are more female planners now than previously. Parallel to this, male planners and engineers are taking on more responsibility for the home, such as taking children to preschool. These shifts in the composition of the planning community seem to have made an impact upon the discursive space that forms planning practice. The professionals in my study do include what have typically been called 'female' experiences in their interpretations of sustainable mobility and users of the transport system. There is not only an awareness that having children affects mobility practices, but the professionals even have their own experiences of making 'care trips'. So, representations do matter. When professionals have a wide set of experiences, their understanding of space, and probably planning practices too, will change. Therefore, it is important that employers continue working with conscious recruitment and improving the working environment so that everyone will feel welcome. The same goes for educators. This includes integrating a gender perspective throughout the education system and working life.

However, representation as a method is not unproblematic. Firstly, it puts a lot of pressure on individuals to include the experiences they are thought to embody. For instance, having a 50/50 representation of men and women does not mean that all experiences have the same opportunity to be heard. Secondly, it is impossible for professionals to embody the experiences of all citizens. Gender is only one of several categories that will affect your experiences of travelling. Categories such as ethnicity, gender, sexuality, functionality and poverty will also affect your opportunities to be a mobile subject. It is, of course, impossible to count on professionals embodying all of these diverse experiences. In this chapter, I have explained how, even though both men and women were represented at the municipality I studied, and they shared similar experiences in the interviews, these experiences were imbued with middle-class ideals. Alas, we cannot rely on 'experience' as an inherent force that will solve equality issues. Rather, it is the knowledge base of planners that needs to be broadened. Methods for expanding the knowledge base within planning are scarce. This is particularly true for transport planning, which is known for its focus on quantitative measures and models (Vigar 2017). But an experiment in a research project concerning public transport and coordination that my colleagues and I designed after we identified a common lack of public transport experience among professionals displays promising results (see Henriksson 2017). We asked professionals engaged in the research project to attempt public transport for a two-week trial. Ten 'test riders', all with different responsibilities for urban development (including public transport) in a Swedish municipality, tried public transport for two weeks. They all received a two-week public transport pass that they could use at all times on the local and regional buses and trains. We asked them to use their passes as much as possible, and to perform different kinds of errands with public transport, including service and shopping errands. Among other things, the opportunity to try public transport for a limited period allowed the professionals to get a glimpse of what it means to be less privileged in terms of mobility. The experience gave these professionals and stakeholders a reason to reflect upon not only the quality of the local public transport but also the everyday mobility of citizens. These were seen as valuable insights for the participants, and useful from a professional perspective.

The test-ride experiment relates to how planning theorist Leonie Sandercock (1998) describes the epistemology of planners. She states that planners must make use of knowledge claims about the city and its inhabitants in a way that makes it possible to include multiple interests. In particular, this implies that planners must question their role as experts and 'knowers'. From Sandercock's perspective, it is problematic that the professionals described in this chapter are claiming to speak for 'everyone', when they are in fact speaking from a specific, and quite privileged, position. When relating their professional practice to their private practice, they are reinforcing their own experiences as central. However, I argue that experiments such as the test ride can provide planners with an arena to discuss professional and personal knowledge and what kind of experiences they are lacking today. Expanding the knowledge base of professionals could prove helpful if the goal of transport planning (and similar activities) is to reflect a diverse set of experiences.

APPENDIX

Informants

Figured name	Gender (f/m)	Year of birth	Title/role
The Urban Planning Office			
Hans	m	1957	Parking coordinator
Henrik	m	1974	Traffic planner
Herman	m	1973	Traffic engineer, project leader of the bicycle programme
Hector	m	1983	Traffic engineer
Henry	m	1967	Energy and climate adviser
Hilda	f	1981	Traffic planner
Hanna	f	1978	Office coordinator, previously project manager sustainable mobility
Hannes	m	1971	Project manager sustainable mobility
Hedwig	f	1978	Project manager sustainable mobility (sub.)
Heidi	f	1973	Project manager sustainable mobility
Hillevi	f	1959	Traffic engineer (executive)
Henning	m	1949	Traffic director (executive)
Holger	m	1967	Traffic manager (executive)
Harald	m	1948	City gardener (executive)
The Environmental Office			
Hampus	m	1968	Environmental inspector
Helena	f	1973	Environmental strategist

NOTES

1. The full study is published in Henriksson (2014).
2. The "free choice of school" is a political reform from 1992 allowing parents to freely choose school for their child to attend. This as an alternative to being assigned to the closest school in relation to where you live.
3. The image depicts a woman riding a bike with two children on it. They are all wearing helmets and are dressed in bright colours.
4. That childcare trips are mostly performed by women has been studied and problematized within feminist geography; see, for example, Domosh and Seager (2001).

REFERENCES

Akrich, M. (1992). The description of technical objects. In W. E. Bijker & J. Law (Eds.), *Shaping technology/building society: Studies in sociotechnical change* (pp. 205–224). Cambridge, MA: MIT Press.

Akrich, M. (1995). User representations: Practices, methods and sociology. In A. Rip, T. J. Misa, & J. Schot (Eds.), *The approach of constructive technology assessment*. London: Pinter Publishers.

Bacchi, C. L. (1999). *Women, policy and politics. The construction of policy problems.* London: Sage.

Banister, D. (2008). The sustainable mobility paradigm. *Transport Policy, 15,* 73–80.

Böhm, S., Jones, C., Land, C. & Paterson, M. (eds.), 2006. *Against automobility.* Malden/Oxford/Carlton: Blackwell Publishing.

Bradley, K. (2009). *Just environments: Politicising sustainable urban development.* Dissertation, KTH/Royal Institute of Technology, Stockholm.

Cupples, J., & Ridley, E. (2008). Towards a heterogeneous environmental responsibility: Sustainability and cycling. *Area, 40*(2), 254–264.

Czarniawska, B. (2002). *A tale of three cities: Or the glocalization of city management.* Oxford: Oxford University Press.

Davies, A. R., Fahy, F., & Rau, H. (2014). *Challenging consumption: Pathways to a more sustainable future.* New York: Routledge.

Doan, P. (2011). Why question planning assumptions and practices about queer places? In P. Doan (Ed.), *Queering planning: Challenging heteronormative assumptions and reframing planning practices.* Tallahassee: Florida State University.

Domosh, M., & Seager, J. (2001). *Putting women in place: Feminist geographies make sense of the world.* New York: Guilford.

Feinstein, S., & Servon, L. (2005). Introduction: The intersection of gender and planning. In S. Feinstein & L. Servon (Eds.), *Gender and planning: A reader.* New Brunswick: Rutgers University Press.

Friberg, T. (1993). *Everyday life: Women's adaptive strategies in time and space.* Stockholm: Swedish Council for Building Research.

Friberg, T. (2006). Towards a gender conscious counter-discourse in comprehensive physical planning. *GeoJournal, 65,* 275–285.

Friman, M., Larhult, L., & Gärling, T. (2013). An analysis of soft transport policy measures implemented in Sweden to reduce private car use. *Transportation, 40,* 109–129.

Grange, K. (2010). Mellan skrå och profession. Om de svenska arkitekt- och ingenjörsutbildningarnas framväxt och hur ett dominerande kunskapsideal har tagit form [Between guild and profession. On the emergence of Swedish educational programmes for architects and engineers and how a dominant ideal of knowledge have taken shape]. *FORMakademisk – Research Journal of Design and Design Education, 3*(2), 26–38.

Greed, C. (1994). *Women and planning.* London: Routledge.

Greed, C. (Ed.). (1999). *Social town planning.* London: Routledge.

Heisserer, B. (2013). *Curbing the consumption of distance? A practice-theoretical investigation of an employer-based mobility management initiative to promote more sustainable commuting.* Dissertation, National University of Ireland, Galway.

Henriksson, M. (2014). *Att resa rätt är stort, att resa fritt är större. Kommunala planerares föreställningar om hållbara resor.* Dissertation, Linköping University Press, Linköping.

Henriksson, M. (2017). Lived experience as a source of knowledge when planning for public transit: The case of region Västra Götaland. Proceedings. *Regions in Transitions,* NoRSA, 9–10 March 2017, Karlstad, Sweden.

Hjorthol, R. (2000). Same city different options: An analysis of the work trips of married couples in the metropolitan area of Oslo. *Journal of Transport Geography, 8,* 213–220.

Holmberg, T., & Ideland, M. (2009). Transgenic silences: The rhetoric of comparisons and the construction of transgenic mice as 'ordinary treasures. *Biosocieties, 4*(2), 165–181.

Holth, L., & Mellström, U. (2011). Revisiting engineering, masculinity and technology studies: Old structures with new openings. *International Journal of Gender, Science and Technology, 3*(2), 313–329.

Hrelja, R., & Antonsson, H. (2012). Handling user needs: Methods for knowledge creation in Swedish transport planning. *European Transport Research Review, 4*(3), 115–123.

Hrelja, R., Isaksson, K., & Richardson, T. (2013). Choosing conflicts on the road to sustainable mobility: A risky strategy for breaking path dependency in urban policy making. *Transportation Research Part A: Policy and Practice, 49,* 195–205.

Koglin, T. (2013). *Vélomobility: A critical analysis of planning and space.* Dissertation, Lunds Universitet, Lund.

Koskela, H., & Pain, R. (2000). Revisiting fear and place: Women's fear of attack and the built environment. *Geoforum, 31*(2), 269–280.

Larsson, A., & Jalakas, A. (2008). *Jämställdhet nästa! Samhällsplanering ur ett genusperspektiv.* Stockholm: SNS förslag.

Levy, K. (2013). Travel choice reframed: 'Deep distribution' and gender in urban transport. *Environment and Urbanization, 25,* 47–63.

Marsden, G., & Rye, T. (2010). The governance of transport and climate change. *Journal of Transport Geography, 18*(6), 669–678.

Oudshoorn, N., & Pinch, T. (2008). User-technology relationships: Some recent developments. In E. J. Hackett, O. Amsterdamska, M. Lynch, & J. Wajcman (Eds.), *The handbook of science and technology studies* (pp. 541–565). Cambridge, MA: MIT Press.

Oudshoorn, N., Rommes, E., & Stienstra, M. (2004). Configuring the gendered user as everybody: Gender and design cultures in information and communication technologies. *Science, Technology & Human Values, 29*(1), 30–63.

Paterson, M. (2008). *Automobile politics: Ecology and cultural political economy.* New York: Cambridge University Press.

Paterson, M., & Stripple, J. (2010). My space: Governing individuals' carbon emissions. *Environment and Planning D: Society and Space, 28,* 341–362.

Sandercock, L. (1998). *Towards cosmopolis.* Chichester: Wiley.

Sandercock, L., & Forsyth, A. (2005). A gender agenda: New direction for planning theory. In S. S. Fainstein & L. J. Servon (Eds.), *Gender and planning: A reader.* New Brunswick: Rutgers University Press.

Shove, E. (2010). Beyond the ABC: Climate change policy and theories of social change. *Environment and Planning A, 42,* 1273–1285.

Shove, E., Pantzar, M., & Watson, M. (2012). *The dynamics of social practice.* London: Sage.

Snyder, M. G. (1995). Feminist theory and planning theory: Lessons from feminist epistemologies. *Berkeley Planning Journal, 10*(1), 91–106.

Star, S. L. (1991). Power, technology and the phenomenology of conventions: On being allergic to onions. In J. Law (Ed.), *A sociology of monsters: Essays on power, technology and domination* (pp. 26–55). London/New York: Routledge.

Törrönen, J. (2002). Semiotic theory on qualitative interviewing using stimulus texts. *Qualitative Research, 2*(3), 343–362.

Uteng, T. P., & Cresswell, T. (Eds.). (2008). *Gendered mobilities.* Aldershot: Ashgate.

Valentine, G. (1989). The geography of women's fear. *Area, 21*(4), 385–390.

Vigar, G. (2017). The four knowledges of transport planning: Enacting a more communicative, trans-disciplinary policy and decision-making. *Transport Policy, 58,* 39–45.

Woolgar, S. (1991). Configuring the user: The case of usability trials. In J. Law (Ed.), *A sociology of monsters: Essays on power, technology and domination* (pp. 57–99). London: Routledge.

CHAPTER 9

Towards an Intersectional Approach to Men, Masculinities and (Un)sustainable Mobility: The Case of Cycling and Modal Conflicts

Dag Balkmar

INTRODUCTION

Current policies for the Swedish transport system set out to create sustainable, safe, gender-equal and inclusive transport solutions for its citizens (Prop 2008/09:93). Today, one-third of all CO_2 emissions come from road traffic, and such emissions are gendered (Kronsell et al. 2015). Previous studies have shown that men's mobility is more car dependent than women's (Carlsson-Kanyama et al. 1999; Gil Solá 2013; Polk 2009). Compared to women, men travel 45% further by car per day (Kronsell et al. 2015). To ensure transport options that are both gender equal and sustainable, there is a need for increased transport planning for sustainable modes of travel such as cycling, walking and public transport rather than more access to, and usage of, cars. Part of the obstacle to achieving such

D. Balkmar (✉)
Centre for Feminist Social Studies, School of Humanities, Education and Social Sciences, Örebro University, Örebro, Sweden
e-mail: dag.balkmar@oru.se

© The Author(s) 2019
C. L. Scholten, T. Joelsson (eds.), *Integrating Gender into Transport Planning*, https://doi.org/10.1007/978-3-030-05042-9_9

goals is, as argued by Koglin and Rye (2014), that motorized transport takes up much more city space than cycling and that cycling receives too little attention from transport planners.

Both car travel and the ideas of freedom and movement associated with the car are persistently linked to a masculine domain and masculine identity (Balkmar 2012; Joelsson 2013; Mellström 2004). Changing travel patterns from unsustainable cars to collective means of travel, walking and cycling is not only about changing daily habits. It is also about critically engaging with the many ways in which technologies of movement contribute to the (re)production and contestation of masculinities and men's mobility practices (cf. Kronsell et al. 2015), including adaptation to more 'ecomodern' forms of men and masculinities (Anshelm and Hultman 2014). Here, as I will go on to argue, some men and masculinities are constructed as both part of the solution and part of the problem, in the latter case, related to modal conflicts in cities.

Over the last few years, modal conflicts (e.g. aggressive driving, risk-taking and harassment, including verbal violence and sexism) have been discussed in media reports as an everyday problem for cyclists, drivers and pedestrians in both urban and rural areas in Sweden (Andersson 2015; Hallemar 2014; Kronqvist 2013; Lerner 2016; Ljones 2015; Stigel 2015; Ternebom 2013). Modal conflicts are not only a problem for cities with a low prevalence of cycling, as previous studies have shown, but even cities like Copenhagen, with its world-famous cycling-friendly infrastructure, are troubled by conflicts between cyclists and drivers (Freudendal-Pedersen 2015; Koglin 2013). On the one hand, cycling blogs (Gillinger 2013) have critiqued the media for exaggerating and producing oversimplifications of a polemical nature about how things are (e.g. bike wars). On the other hand, in his study of cycling in Stockholm and Copenhagen, Koglin notes that while a majority of the Copenhagen cyclists found it safe to cycle (67%), the majority of the Stockholm cyclists felt it was not safe (58%) (Koglin 2013: 148). While cities such as Stockholm aim to become as 'bike friendly' as Copenhagen, there is a long way to go to reach this goal: only 7% of Stockholm's residents cycle to work or school daily; the equivalent for Copenhagen is 38% (Pirttisalo 2015, 41% in 2017, http://www.cycling-embassy.dk/2017/07/04/copenhagen-city-cyclists-facts-figures-2017/). Cyclists' experiences of safety and risk vary between cities, including what forms of modal conflict are considered annoying, for example, with car drivers or other cyclists (Koglin 2013; Hartley 2017). Hence, it is very likely that the occurrence of such conflicts varies between cities, parts of cities, and urban and rural areas, including between different forms of mobility, including between different cycling categories or between cyclists and pedestrians. Having said this, in order to succeed in creating cities that truly promote cycling, there is also a need to challenge motorized traffic and its normative influence on city planning (Lindkvist Scholten et al. 2018).

What makes modal conflicts important to consider from a planning and gender perspective is that, for cycling to increase, according to the Swedish cycling safety strategy, it needs to be *experienced as safe* (Trafikverket 2014). Bike safety is therefore an important condition for achieving sustainable mobility (Aldred and Crosweller 2015; Jacobsen et al. 2009). As scholars working on gender and mobility have shown in other countries than Sweden, dangerous traffic and fear of violence have major influences on travel patterns, causing women in particular to refrain from dwelling in public space, including cycling (Emond et al. 2009; Heesch et al. 2011). However, even though certain men and certain masculinities play key roles in risk-taking and environmental damage, men and masculinities remains an understudied area in critical debates and studies on men, masculinities and (un)sustainable mobilities (Balkmar and Hearn 2018).

Therefore, my aim in this chapter, through the voices of cyclists themselves, is to discuss cycling promotion and modal conflicts in public space with a particular focus on men, masculinities and transport planning. The goal is not only to contribute to research on Swedish transport politics and the gender domain, but also to highlight some of the consequences the analysis at hand may have for transport planning. Against this background, the chapter draws on three sources of empirical material: (1) online discussions produced by cyclists about cycling, (2) interviews with cyclists and cyclist advocacies about cycling and how to promote cycling, and (3) media material on cycling and risk. Although the material upon which I rest my claims is limited and tentative, it was chosen to mirror three areas of conflict that have emerged in contemporary discourses on cycling, the red thread being men, masculinities and (un)sustainable mobilities. These are:

- Men as solution: for example, the promotion of men as sustainable subjects
- Men as problem: for example, risk-taking men as problematic road users
- Men as vulnerable road users: for example, unprotected cyclists subjected to (often male) drivers' violent and aggressive driving practices

The first two examples are chosen to illustrate how men and masculinities can be framed as both *solutions* and *obstacles* to achieving more sustainable mobilities through more cycling. The third example is chosen because it shifts the perspective by demonstrating how cycling implies a particularly vulnerable and conflicting position in the traffic hierarchy. Vulnerability, modal inequalities and gendered conflicts are then used as examples to suggest an intersectional approach that considers the role of mobilities in the production of inequalities in public space, moving beyond gender binaries. Hence, these examples encompass different levels and arenas, ranging from spokespersons for cycling-promoting organizations and their narratives on how to best promote cycling at a political level (men as solutions); to individual cyclists and media representations of male cyclists' risk-taking practices (men as problem); to individual (male) cyclists' experiences of vulnerability in car-dominated space (men as vulnerable). Together, they exemplify three interlinked areas of conflict in the context of men, masculinities and sustainable mobilities.

In the following, I will provide a background to the chapter, followed by a theoretical framework that will guide the analysis, including a short description of the method and empirical material. After the analysis, the chapter will conclude with a discussion of how an intersectional approach to road conflicts and power may contribute to a more complex analysis of transport politics and planning.

BACKGROUND: MEN, MASCULINITIES AND TRANSPORT POLICY

As a background to this chapter, I will reflect upon some of the key policy writings and reports that focus on gender and transport more generally, and cycling in particular. I will do this as a way of illustrating how men and masculinities have been constructed in Swedish transport politics in the past. Here I draw on Bacchi (2009) and Mouffe (2005), who argue that policy is *done* and shaped in an ongoing conflictual process. Hence, particular policies create particular understandings of what the *problem* is, including its preferred solutions. I refer to gender as context dependent and intersecting with other social hierarchies, such as class, ethnicity and age (Hanson 2010). Hence, gender is assumed to be not a fixed or innate source of difference, but embodied, dynamic and socially constructed (Butler 1990).

Sweden recognizes the importance of considering gender aspects in transport policy and since 2001 it has implemented explicit gender equality goals in its transport policy (Svedberg 2013; Kronsell et al. 2015). The overall aim of Swedish transport policy is to ensure efficient and long-term sustainable transport for business and citizens all over the country (Prop. 2008/09:93). This target consists of two parts, in which the so-called *functional target* seeks to create accessibility in terms of design, function and usability for its users. The functional target also states that 'the transport system should respond equally to women's and men's transport needs'. The second part is called *the considerations target*, and focuses on increased accessibility, along with improved health, road safety and environmental performance. It follows from the considerations target that no one should be killed or seriously injured in Swedish traffic and that the design, function and utilization of the transport system are to be adapted to meet this aim.

Even though the male norm has been problematized in previous studies, an explicit focus on men and masculinities has been rare in studies of transport and transport policy. As Kronsell et al. (2015) note, male norms are still dominant in the transport sector even in 'gender-equal' Sweden. Masculinity is 'the accepted norm, deeply embedded in the transport sector' (Kronsell et al. 2015: 709). For example, male norms may find expressions in men being overrepresented on decision-making bodies. Kronsell et al. (2015: 704) argue that '[i]f the norms of the transport sector are masculine, they are likely embedded its policy institutions and likely to constrain activities, such as decision making and planning, in accordance with these dominant masculine norms'. Following their argument, one could argue that part of the problem is how masculinity becomes implicit in decision-making and planning, rather than explicitly recognized and problematized. For example, while the functional target formulation explicitly includes gender equality in the establishing of its goal, the consideration target addresses gender differences related to risk-taking in the general policy text, as an issue to be taken into account in the future (Prop. 2008/09:93). In Sweden, men make up 88% of those prosecuted for traffic violations, but do 70% of the driving (SCB 2016; Trafikanalys 2015). Even though Swedish policy writings on traffic safety explicitly address risk-taking practices as a gendered problem where men are overrepresented, it is rarely problematized further as a dangerous and violent way of performing everyday masculinities (Joelsson 2013). Hence, while men's risk-taking is emphasized as a problem for reaching traffic-safety goals, there is little further discussion of the ways in which these gendered problems should be addressed or how they could be taken into account in, for example, transport planning.

Research reports that explicitly deal with the problem of men's violence in public space do this from the perspective of the national gender equality goal that focuses on men's violence towards women. This goal is often translated into safety measures in public transport space, that is, street lighting and other measures for 'eliminating risks and fears of being subjected to gender-based violence/crime in relation to transport' and to obviate negative retroaction on mobility (Faith-Ell and Levin 2013: 57). Such measures are, of course, extremely important for improving safety in public space, but they do not have as their primary focus issues related to the ways in which violent mobilities are linked with men, masculinities and automobility.

In reports on cycling and safety, such as the governmental cycling investigation from 2012 (SOU 2012: 70), and the Cycling Safety Strategy (Trafikverket 2014), safety is at the centre of attention, yet discussed in non-gendered ways. Having said this, it should be mentioned that the former (SOU 2012: 70) does discuss improvements in cycling safety and accessibility as beneficial for reaching gender equality goals, as stated in the beneficial targets. Much as feminist critiques of gender mainstreaming as a strategy have pointed out, what is often lacking are explicit formulations of *how* this can be done and by *whom*, which is not further discussed (Alnebratt and Rönnblom 2016). The Cycling Safety Strategy 2014–2020 (Trafikverket 2014), which is a joint strategy formulated by researchers, cycling experts, cycling organizations, municipalities, insurance companies and the Swedish traffic organization, clearly states that, until now, cyclists have had to conform to a road transport system that is not made for them, where space has been given to cars. A greater responsibility has been placed on cyclists to ensure their own safety, compared to motorists, who have had the system applied in terms of their needs and safety requirements. As argued by Lindkvist Scholten et al. (2018: 4), to achieve political goals on more sustainable mobilities such as cycling, there is a need for clear political leadership on this matter, strategic planning that integrates sustainable transport such as cycling early on, and a planning practice with the mandate to challenge car-normative infrastructure by transferring space from motorized traffic to bikes.

While the cycling safety strategy primarily argues for more cycle-friendly infrastructure, issues related to interplay, norms and behaviour are also brought up as important for achieving more cycling: 'The work on raising the status of cycling also includes influencing the behaviour of motorists, pedestrians and cyclists in various ways, along with their interaction with each other in traffic. However, undesirable behaviour is not primarily a problem of knowledge but rather a matter of norms or standards' (Trafikverket 2014: 12). Issues related to men, masculinities and risk-taking[1] do not form part of this discussion, nor how experiences of (un) safety may be gendered or how such norms should be changed. Hence, while power relations can be addressed and problematized in relation to access to public space, conflicting power relations in the context of *doing traffic* (including problematizing and gendering the 'doers' and victims of violent mobilities) are generally not further examined.

In addition, while gender and gender equality remain the dominant frame in Swedish transport policy, concepts of multiple inequalities and intersectionality have also begun to be used in more recent transport research, suggesting that more complex analytical lenses are being called for and developed (Dahl 2014; Faith-Ell and Levin 2013; Henriksson 2014). Even though cycling can be conceived of as a truly democratic and accessible form of mobility, history shows that this has not been the case in practice. The bicycle has historically played an important role in class-based as well as gender-based struggles for social change. For example, in 1895 Frances Willard, an American suffragist who learned to ride a bicycle, recognized several parallels between riding a bike as a woman and the wider struggle for women's rights (Hanson 2010; Willard 1997). A more recent case is presented in a study by Steinbach et al. (2011) on how gendered, ethnic and class identities can shape becoming a cyclist or not. One of its main findings concerned the congruence between being someone who cycles, and an entire social identity bound up with class, ethnicity, values and aspirations. Lugo argues that, in struggles to make space for cycling on public roads, it is not only distinctions relating to whether you are a driver or a cyclist that matter, but also issues around race, class and gender (Lugo 2016: 184). Hence, these aspects clearly point to a need for more progressive transport planning, where intersectionality can be a fruitful approach when analysing, for example, who may become a cyclist and identify with cycling, including who is taken to represent cyclists, or not, when cycling is promoted.

AUTOMOBILITY AND VIOLENT MOBILITIES

Automobility can be conceived of as a regime, as one of the principal socio-technical political institutions and practices that organize and shape spatial movement (Böhm et al. 2006). Böhm et al. (2006) argue that the following dimensions combine to legitimize its dominance: its centrality as an ideological and discursive formation; its capacity to embody ideals of privacy, freedom, movement, progress and autonomy; and automobility as phenomenology and ways of experiencing the world. Combined, its principal technical artefacts, such as roads and cars, are being legitimized through 'collective human agency in the production of automobility' (Böhm et al. 2006: 5). This in turn, they argue, tends to make it appear natural and non-political.

However, a key part of their argument is that by highlighting its inherent antagonisms, such a regime may also be challenged and undermined. One way of doing this is by taking seriously the critical discourses produced by cyclists and activists about what cycling in the current automobility regime can be like (Balkmar and Summerton 2017). Hence, as discussed in the introduction to this book, the automobility regime is a concept that fits well with Mouffe's understanding of hegemony as reproduced through practices that in turn make it appear natural—nevertheless, with a 'present possibility of antagonism' (Mouffe online).

One way of revealing its inherent antagonisms is through the perspective of violence. Joelsson (2013) argues that automobility can be theorized as a violent regime, highlighting how traffic accidents, collisions and risk-taking with motor vehicles in public space can be understood as a kind of violence. She argues that risk-taking, typically a male-dominated practice, can be 'understood as part of sociocultural scripts of violence, with gendered associations and implications' (Joelsson 2013: 218). From a cyclist's perspective, aggressive driving, risk-taking and harassment of cyclists can exemplify this (Balkmar 2018). Furthermore, violent mobilities include not only directly violent practices, but also so-called near-misses (cf. Aldred and Crosweller 2015), that is, incidents that cyclists may experience as frightening, annoying and/or violent acts. Therefore, I draw on Jeff Hearn and Wendy Parkin's (2001: 18) broad understanding of violence, defined as actions, structures and events that violate or cause violation or are considered as violating. Hence, violent mobility is not only about intentionally exposing someone to potential risks when committing a traffic offence, in fact, automobility can be viewed as a 'violent regime' that puts cyclists at risk in car-dominated contexts (Balkmar 2018; Joelsson 2013).

METHOD AND MATERIALS

The core material for this analysis arises from a study on cyclists' experiences of violence and harassment in traffic space based on ten in-depth interviews with cyclists, and online ethnographic material (see Balkmar 2018). The in-depth interviews encompass five women and five men and focus on their cycling experiences, mostly in Stockholm, while three of the men were based in the middle-sized city of Linköping. Both Stockholm and Linköping aspire to be perceived as cycling-friendly cities, which makes them relevant contexts for this study. However, it is primarily Stockholm that informants and media representations associate with considerable conflict between cyclists and other road users. As well as cycling on an everyday basis, four of the men and two of the women are experienced leisure cyclists. Four of the cyclists (two of the women and two of the men) also identify themselves as bike advocates/activists. Most interviews took place in 2014–2015; one was conducted in 2007. At the time of the interviews, informants were between 23 and 55 years of age, the majority from the ethnic Swedish population. They were conducted in the informant's home/work, or via Skype/telephone and took between 1 and 2.5 hours. Informants were asked to talk about specific occasions that they consider as having been particularly scary, risky or violent, including describing what happened and how they dealt with the situation.

For this chapter, I also draw on a case study carried out in three Swedish online cycling forums: *Happymtb.org*, which is Sweden's largest cycling community (at the time of the study, 270,000 unique visitors/month; since 2016 named *Happyride.se*), *Cykelforum.se* (data on members unavailable online) and *Funbeat.se* (100,000 unique visitors/week). The data chosen for this particular chapter was generated by a person presenting herself as a researcher in the above-mentioned cycling forums in November 2014, and asking members to comment on the questions: 'How are cyclists treated in Sweden?' and 'Are cyclists treated differently?', a call that generated 142 responses. These online public discussions offer rich material for studying cyclists' meaning-making, in this case especially their experiences of risk, conflict, pleasure and vulnerability in everyday cycling (for a more extended discussion, see Balkmar 2018).

During the analytical process, I noted themes of importance, such as talk about unequal power relations, and experiences of demonstrations of power, vulnerability and violence. The interview questions in the next section, which is based on a study of Swedish cycling activism (see Balkmar and Summerton 2017), revolved around the activists'/advocates' approaches and activities, their views of cycling advocacy and activism more generally, and their visions for the future of cycling.

ANALYSIS

Men as Solution

While contemporary cycling in Sweden is fairly gender equal in terms of uptake, it is also intersectional in complex ways. For example, cyclists in Stockholm are usually described as middle-class, well-educated and 'time-aware' commuters (Emanuel 2012). However, over time, the bicycle and cycling have been gendered, classed and aged in many different ways. When it was first invented, the bicycle was a vehicle for rich white men (1890), over time it became a vehicle for workers and the middle classes (1930–40), and then more and more became a sign of poverty with the arrival of the car (1940–50) (Emanuel 2012). Today, the bicycle is described as a trendy artefact for the middle and upper classes, even though there are frequent users in all classes (Trivector 2014). According to Trivector (2014), symbolically, the bicycle has shifted from a vehicle associated with children's play and for the poor to an iconic design artefact for sustainable, healthy, 'green', sporty, 'creative', middle-class, urban mobile subjects. However, while this might be true for Swedish middle-class subjects living in urban parts of Sweden, it should be emphasized that gendered, ethnic and class-related dynamics influence who may identify themselves as a cyclist or not (see Steinbach et al. 2011, pedalista.org).

Despite the variety of cyclists and identities in everyday cycling, classed and gendered norms are being reproduced in pro-cycling discourses. In this example, which is discussed at greater length elsewhere (Balkmar and Summerton 2017), the head of the Swedish national cycling advocacy organization is talking about role models and spokespersons for cyclists in a particularly gendered and classed way:

> [I]t has helped a lot that the educated, masculine middle class in large cities and regions have started to ride bikes, because then you have editorial pages writing about cycling and popular musicians ride bikes, CEOs ride bikes – I feel strongly that this has meant that the cycling movement and cycling activism can be expressed in different ways, which I think is extremely positive.

Men in positions of power and prestige are here stressed as particularly important in political lobbying for cycling. The director of Sweden's largest cycling advocacy organization emphasizes the political and symbolic importance of cyclists who belong to the 'educated, masculine middle class', a category thereby identified as particularly influential for promoting cycling. He refers to these cyclists as a resource for advocacy work; men add to the plurality of bicycle advocacy and activism specifically with regards to their (privileged) class, gender and ethnic positions (Balkmar and Summerton 2017: 162).

The construction of some men and masculinities as important for the promotion of cycling can also be exemplified in studies on how planners imagine sustainable mobility. Malin Henriksson (2014) has examined how planners imagine sustainable mobility and, more particularly, who the sustainable traveller is; see also Henriksson in this collection. Primarily, the planners constructed men in suits into role models for sustainable mobility, thereby also constructing the cyclist as male. For example, images picturing bike-riding men in suits were considered very positive, as they made cycling a cool thing to do rather than a low-status alternative (Henriksson 2014). As is common when cycling is promoted, the planners did not want to talk about how cheap cycling is compared to the car; this was imagined as making cycling low status and would make it harder to attract men (Henriksson 2014). Middle-aged men in suits are associated with purchasing power, economic power and successful careers and they were constructed as role models, which in turn made those already on bikes or using buses less visible to the planners.

In addition, Henriksson (2014) notes that planners did not consider all cyclists to be equally important for reducing carbon footprints. While several municipalities in Sweden arrange bicycle-training courses for women with immigrant backgrounds, as an explicit way of empowering a marginalized group through increasing their mobility capacity, this was not an activity that was considered to reduce their carbon footprint (Balkmar and Henriksson 2016). Hence, environmental goals were not considered to be met through these courses (which in turn builds on the idea that social equality can be separated from environmental equality, as argued by Aslam 2015). It also exemplifies, as Henriksson (2014) argues, how middle-class values tend to become the norm in the context of promoting sustainable mobilities.

In conclusion, while cycling in Sweden is a fairly gender-equal practice (although it is aged and urban), by underscoring men's symbolic and political importance as resources for social change, as these examples suggest, pro-cycling discourses may not only reinforce cycling as primarily about white, middle-class men's lifestyle choices (Golub et al. 2016), but this example also points to the risk of rendering invisible cyclists from ethnic minorities, including many women, as agents for change (Balkmar and Summerton 2017: 162).

Men as Problem

Men are not only constructed as solutions to contemporary sustainability problems; some men are also considered to contribute to conflicts in urban traffic. It has been suggested that cars can be used as an aggressive means of achieving mobility at the expense of others—a form of risk-taking that is closely linked to masculinity, automobility and space (Redshaw 2008: 121). Similar issues are also related to stereotypes building on intersections of masculinity, class, age, competitiveness and cycling. Reports about so-called hatred against cyclists are related in media discussions to the rise of the MAMIL (Middle-Aged Men in Lycra), a particularly careless and ruthless kind of male cyclist (Andersson 2015; Hallemar 2014; Lerner 2016). This is a media stereotype that is predominantly related to urban cyclists, a figure that journalist Ola Andersson (2015) critiques as 'the new kind of cyclists' in Stockholm:

> They train hard, are unscrupulous and armed for street fighting. They are MAMILS: Middle-Aged Men in Lycra, spurred on by their struggle for the environment. (Andersson 2015)

Their desperate rampages are, according to this journalist, justified by their view of themselves 'as noble knights on a daily crusade against mass motoring, considering themselves to be the shock-troops for a future when the car is abolished'. Journalist Somar Al Naher goes on to argue that, in order to make cycling safer, not only planning or helmet use matters, but the ways in which 'male cyclists in lycra shorts' behave:

> For more people to want to commute by bike and feel safe, cyclists themselves must also change their behaviour in traffic. This applies primarily to the cycling men in close-fitting lycra who believe they are in the Tour de France. The ones who ignore traffic rules, red lights and almost run over baby carriages at pedestrian crossings. (Al Naher 2017: n.p.)

Interviewees also referred to very fit male cyclists as particularly dangerous road users:

[It's like the] Tour de France, they pass to the right and to the left, they pass everyone, 35-year-olds wearing lycra. They try to exercise to work, but you can't, you have to exercise somewhere else. (Elinor, Stockholm).

According to this informant, and the media writing on the subject exemplified above, fast male cyclists are considered a general risk to slower cyclists and pedestrians. With its associations with the Tour de France, cycling is related to a typically masculine competitive practice, primarily engaged in by men, potentially excluding women. The Tour de France racer exemplifies the ability to cycle fast and take risks, an image that does not include caring for others (cf. Joelsson 2013: 189). Rachel Aldred (2013: 256) discusses sporty cycling bodies as being considered particularly problematic in the sense of doing exercise in the 'wrong place'. When faster cyclists confuse the boundaries between cycling as everyday activity and sport, in this case the lycra-wearing man, cycling is not only turned into exercise and competition but also comes to exemplify the performance of careless masculinity (Aldred 2013: 256; Joelsson 2013).

In conclusion, the image of the 'Tour de France' cyclist can be related to the category of high-status cyclists that the planners considered to be their ideal in Henriksson's (2014) study. Only here, 'the suit' has changed into his lycra outfit and comes to articulate an aggressive form of mobility by using his expensive sports bike without consideration for other road users' safety or traffic laws. Rather than contributing to more sustainable mobility, the masculine sporty cyclist manipulates and subverts the dominant rhythm of traffic (Spinney 2010) and, by doing so, is considered to generate anger and risk for other road users in potentially violent ways (Hearn and Parkin 2001). Rather than being constructed as a solution to the present transport situation, these male commuters become potential problems in the quest to get more people to travel by bike.

Vulnerable Men

Men are not only constructed as solutions or problems in current cycling discourses; cycling also encompasses vulnerabilities that are not gendered in any clear-cut way. In the following, I introduce how cyclists in interviews and online communities discuss their situation as cyclists in Sweden.

Even though the material consists of various examples of how cyclists (presumably men) share amongst themselves their cycling experiences, the data also consists of discussions about how to improve the transport system and built environment for safer cycling. However, the central topic in the online discussions chosen here is how the contributors discuss vulnerability.

In general, the online discussions create an image of cyclists as being vulnerable in the traffic hierarchy, as exposed to other road users' mishaps, facing a lack of respect and lack of safety. The problem, as many contributors see it, is the general attitude that cyclists don't belong on Swedish roads.

> My experiences are that cyclists should not be in the way, motorists should be able to pass cyclists without any delay, regardless of the traffic situation. You sort of forget that it is a vehicle that has the same right to travel on the road as every other vehicle. This is a generalization and does not apply to all, but to many. (Happymtb.org)

According to informants and online material, these experiences exemplify the fact that cyclists are less valued than car drivers and are still not treated as legitimate road users. In line with previous research from far less 'bike-friendly' contexts than Sweden (e.g. the USA, McCarthy 2011; the UK, Aldred and Crosweller 2015), the informants and contributors to online forums talk about incidents directed against them specifically as cyclists:

> It's the motorists' idea that roads are for them and cyclists are intruders that lies behind the hailing, honking, spraying of windscreen fluid and all the rest. (Happymtb.org)

Actions like these may in themselves be experienced as frightening and violent, even though not resulting in direct physical injury (Lugo 2013). Threatening situations, as the following contributor suggests, also affect a cyclist's sense of self:

> The problem is those times when you experience threatening situations, well it gets to you really hard because I am an unprotected road user and because I feel the lack of respect. (Happymtb.org)

Such lack of respect may over time, as in the following excerpt, come to be considered as normality:

> To some extent I have become tame, have almost become used to it. In the beginning, I got upset but after a while I didn't have the stamina anymore. You've faded to that feeling sort of, like, it's become more like that's the way it is. (David, Stockholm)

Following this excerpt, it seems as though the cycling position, at least temporarily and to some extent, comes to undermine dominant ideals of men and masculinity; namely, that men should be able to defend themselves when disrespected or harassed (Balkmar 2018; Hearn and Parkin 2001). However, in most of the material upon which this chapter is based, both male and female cyclists refer to how they have become 'tame' and come to accept the situation. Avoidance of being provoked by drivers and refraining from responding, by shouting or gesturing back, was stressed as a safety strategy by several informants (Balkmar 2018). This also exemplifies how the automobility regime demands vulnerable road users' consent to discipline themselves in order to stay 'safe' (Jain 2005). This, in turn, shows how automobility needs to be understood as a violent regime for vulnerable road users, with gendered implications (Joelsson 2013). It follows that in order to succeed in meeting policy goals on increased cycling, there is a need for political leadership and transport planners to critically consider the conditions under which cycling take place, and to clearly prioritize cycling in transport planning and space.

COMPARATIVE REFLECTIONS AND CONCLUSIONS

In this final section, I will use the three areas of conflict briefly outlined above (men as solution, men as problem and men as vulnerable) to reflect upon transport politics, the gender domain, intersectionality and transport planning more particularly.

First, from the perspective of the cyclists interviewed for this study, cycling can be experienced as both truly pleasurable and at times outright violent. The gendered implications of the latter are several. One aspect to consider is how the automobility regime produces uncaring and violent configurations of men and masculinity (Balkmar 2018; Joelsson 2013). Control over motorized movement traditionally implies masculinity, maturity and the ability to control others (Balkmar 2018). Automobility makes it possible for some drivers to violently perform their alleged right to the road, practices that typically reproduce traditional scripts associated with men, masculinity and power in public space (Joelsson 2013). Such practices are problematic from both gender equality and sustainability perspectives because they tend to reinstate the motor car as the norm in the traffic hierarchy, reproducing cycling as a marginalized and thus a less accessible means of (sustainable) mobility. From this perspective, the analysis above is indicative of effects that follow from transport planning that still emphasizes motorized transport rather than more sustainable means of movement.

Second, modal conflicts relate not only to constructions of masculinity or individual men's violent driving practices, but to issues relating to planning, mobility and power. From a cyclist's perspective, mobility itself can be interpreted as an axis of inequality. Salazar (2014: 60) suggests thinking of mobility as a 'key difference- and otherness-producing machine, involving significant inequalities of speed, risk, rights, and status'. In Sweden, as well as elsewhere, this can be seen in how traffic planning has been shaped by a legacy in which traffic engineers and urban planners have marginalized cycling while prioritizing the motorist and the car (Emanuel 2012; Koglin 2013). Different mobilities, such as cycling, driving and walking, all have their different rhythms and speeds, and are differently situated in relation to power and transport hierarchies (Freudendal-Pedersen and Cuzzocrea 2015; Spinney 2010). From this, it follows that a shift from cars to more sustainable mobility demands related shifts in the context of mobility hierarchies and planning practices more generally, and cars in particular.

Third, from a political point of view, the vulnerable cyclist position may also come to generate potentially transgressive perspectives, as in exposing antagonisms inherent to the current automobility regime. Here I am especially thinking of the political potential of online communities, as when cyclists use online spaces to share, theorize and reflect upon what it means to be a cyclist in car-dominated areas. In fact, one could assume several possible responses to road conflicts: as normalized and over time taken-for-granted normality, or, as discussed elsewhere, spurring cyclists into more confrontational responses and political activism (Balkmar and Summerton 2017). This latter strategy would be concerned with the transformation of a transport system that is viewed by some cyclists as creating conflicts, unsafe conditions, and so-called hatred against cyclists, as compared to the inclusive, bike-friendly urban space that it could be (Balkmar and Summerton 2017).

Fourth, cycling also offers a way for some men to adapt to more 'eco-modern' forms of masculinity. While this includes how questions of sustainable mobility may become more integrated into the concept of masculinity, mobility practices and modernity, this area also, as discussed, entails potential conflicts. In the analysis, I have noted how privileged men and certain masculinities could be framed as role models (as well as problems, depending on whether it is sustainability or traffic safety that is emphasized) in discussions of sustainable mobilities. If cycling promotion and its imagined role models are limited to middle-class men with successful careers, as problematized above, the male norm in the Swedish transport sector remains unchallenged.

Given the discussions above, one way forward would be an intersectional approach, namely by asking *which* categories are made important, and *which are not* (Kaijser 2014: 45, 46). An intersectional approach could offer useful tools for transport planners to make visible, and thereby also potentially address, such un-reflected reproductions of middle-class male norms in contexts such as (un)sustainable mobility, cycling promotion and transport planning. In addition, an intersectional approach also opens up space for transport planners to ask questions about how differences related to class, age, race, ethnicity, gender, disability, body form and sexuality may influence how safety and risk are experienced while (cycling) bodies are in motion (Lee 2016; Nixon 2014). By further considering how such differences may shape everyday cycling experiences, transport politics and planning may become better able to meet the demand for the safer and more sustainable and inclusive transport systems that many cities are striving to achieve.

NOTE

1. However, in recent reports on cyclist safety, men have been singled out as vulnerable road users, and as being *at risk* to a much greater extent than women. A study from the Swedish National Road and Transport Research Institute found that "[f]atalities of males account for two-thirds of the cases" (Ekström and Linder 2017:35).

REFERENCES

Aldred, R. (2013). Incompetent or too competent? Negotiating everyday cycling identities in a motor dominated society. *Mobilities, 8*(2), 252–271.

Aldred, R., & Crosweller, S. (2015). Investigating the rates and impacts of near misses and related incidents among UK cyclists. *Journal of Transport and Health, 2*(3), 379–393.

Al Naher, S. (2017). Myror är smartare än cyklister [Ants are smarter than cyclists]. *Aftonbladet*, March 26. http://www.aftonbladet.se/ledare/a/dA9m1/myror-ar-smartare-an-cyklister

Alnebratt, K., & Rönnblom, M. (2016). *Feminism som byråkrati: jämställdhetsintegrering som strategi* [Feminism as bureaucracy: Gender mainstreaming as strategy]. Stockholm: Leopard.

Andersson, O. (2015). Så blev män i lycra aggressiva cykelkrigare [This is how men in lycra became aggressive cyclist warriors]. *Dagens Nyheter*, June 28.

Anshelm, J., & Hultman, M. (2014). A Green Fatwā? Climate change as a threat to the masculinity of industrial modernity. *NORMA: The International Journal for Masculinity Studies, 9*(2), 84–96.

Aslam, A. (2015). *Vuxna cyklisters cykelanvändning efter genomförd cykelkurs, en explorativ undersökning om effekten av utbildning på cykelanvändning* [Adult cyclists bicycle uptake after fulfilling cycle course]. Examensarbete inom trafik och transportplanering, Stockholm: KTH.

Bacchi, C. (2009). *Analysing policy: What's the problem represented to be?* Frenchs Forest: Pearson.

Balkmar, D. (2012). *On men and cars: An ethnographic study of gendered, risky and dangerous relations*. Dissertation, Linköping University.

Balkmar, D. (2018). Violent mobilities: Men, masculinities and road conflicts in Sweden. *Mobilities*. https://doi.org/10.1080/17450101.2018.1500096.

Balkmar, D., & Henriksson, M. (2016). *Transportforum*, Mobilitet på lika villkor? Om jämlikhet och makt i transportpolitiken [Mobility on equal terms? On equality and power in transport politics]. Presentation Transportforum, Linköping, 13 January.

Balkmar, D., & Summerton, J. (2017). Contested mobilities: Politics, strategies and visions in Swedish bicycle activism. *Applied Mobilities*. https://doi.org/10.1080/23800127.2017.1293910.

Balkmar, D., & Hearn, J. (2018). Men, automobility, movements, and the environment: Imagining (un)sustainable, automated transport futures. In J. Hearn, E. Vasquez del Aguila, & M. Blagojević (Eds.), *Unsustainable institutions of men: Transnational dispersed centres, gender power, contradictions* (pp. 225–254). Abingdon/New York: Routledge.

Böhm, S., Jones, C., Land, C., & Paterson, M. (2006). *Against automobility*. Malden: Blackwell.

Butler, J. (1990). *Gender trouble: Feminism and the subversion of identity.* New York: Routledge.

Carlsson-Kanyama, A., Linden, A.-L., & Thelander, Å. (1999). Insights and applications gender differences in environmental impacts from patterns of transportation – A case study from Sweden. *Society & Natural Resources, 12*(4), 355–369.

Dahl, E. (2014). *Om miljöproblemen hänger på mig: individer förhandlar sitt ansvar för miljön* [If handling environmental problems were up to me: Individuals negotiate their environmental responsibility]. Dissertation, Linköping University.

Ekström, C., & Linder, A. (2017). Fatally injured cyclists in Sweden 2005–2015. Analysis of accident circumstances, injuries and suggestions for safety improvements. Report, VTI notat 5A-2017.

Emanuel, M. (2012). *Trafikslag på undantag: cykeltrafiken i Stockholm 1930–1980* [Excluded through planning: bicycle traffic in Stockholm 1930–1980]. Dissertation, Kungl. tekniska högskolan.

Emond, C., Tang, W., & Handy, S. (2009). Explaining gender differences in bicycling behaviour. *Transportation Research Record: Journal of the Transportation Research Board, 2125*(1), 16–25. https://doi.org/10.3141/2125-03.

Faith-Ell, C., & Levin, L. (2013). *Kön i trafiken: jämställdhet i kommunal transportplanering.* [Gender in traffic: Gender equality in municipality transport planning]. Stockholm: Sveriges kommuner och landsting.

Freudendal-Pedersen, M. (2015). Cyclists as part of city's organism: Structural stories on cycling in Copenhagen. *City and Society, 27*(1), 30–50.

Freudendal-Pedersen, M., & Cuzzocrea, V. (2015). Cities and mobilities. *City and Society, 27*(1), 4–8.

Gil Solá, A. (2013). *På väg mot jämställda arbetsresor: vardagens mobilitet i förändring och förhandling* [Towards gender equality? Women's and men's commuting under transformation and negotiation]. Dissertation, University of Gothenburg.

Gillinger, C. (2013). *Det är ett krig, I Tell You, Ett Cykelkrig!* [It's a war, I Tell You, a bike war!]. Cyklistbloggen. http://www.cyklistbloggen.se/2013/06/det-ar-ett-krig-i-tell-you-ett-cykelkrig/. Accessed 30 Oct 2017.

Golub, A., Hoffmann, M. L., Lugo, A. E., & Sandoval, G. F. (2016). Introduction: Creating an inclusionary bicycle justice movement. In A. Golub, M. L. Hoffmann, A. E. Lugo, & G. F. Sandoval (Eds.), *Bicycle justice and urban transformation: Biking for all?* (pp. 1–19). Abingdon: Routledge.

Hallemar, D. (2014). Ojämnt krig om stadsrummet [Unequal war for the city space]. *Svenska Dagbladet,* January 25.

Hanson, S. (2010). Gender and mobility: New approaches for informing sustainability. *Gender Place and Culture: A Journal of Feminist Geography, 17*(1), 5–23.

Hartley, K. (2017). Andra cyklister är det som besvärar Malmös cyklister mest [Other cyclists is what bothers the cyclists of Malmö the most]. *Sydsvenskan*, April 17. https://www.sydsvenskan.se/2017-04-17/andra-cyklister-det-som-besvarar-malmos-cyklister-mest

Hearn, J., & Parkin, W. (2001). *Gender, sexuality and violence in organizations*. London: Sage.

Heesch, K. C., Sahlqvist, S., & Garrard, J. (2011). Cyclists' experiences of harassment from motorists: Findings from a survey of cyclists in Queensland, Australia. *Preventive Medicine, 53*(6), 417–420.

Henriksson, M. (2014). *Att resa rätt är stort, att resa fritt är större: kommunala planerares föreställningar om hållbara resor* [Travelling correctly is great, travelling freely is greater – Municipal planners' images of sustainable mobility]. Dissertation, Linköping University.

Jacobsen, P., Racioppi, F., & Rutter, H. (2009). Who owns the roads? How motorised traffic discourages walking and bicycling. *Injury Prevention, 15*(6), 369–373.

Jain, S. (2005). Violent submission: Gendered automobility. *Cultural Critique, 61*, 187–214.

Joelsson, T. (2013). *Space and sensibility: Young men's risk-taking with motor vehicles*. Dissertation, Linköping University.

Kaijser, A. (2014). *Who is marching for Pachamama? An intersectional analysis of environmental struggles in Bolivia under the government of Evo Morales*. Dissertation, Lund University.

Koglin, T. (2013). *Vélomobility – A critical analysis of planning and space*. Dissertation, Lund University.

Koglin, T., & Rye, T. (2014). The marginalisation of bicycling in modernist urban transport planning. *Journal of Transport & Health, 1*, 214–222. https://doi.org/10.1016/j.jth.2014.09.006.

Kronqvist, P. (2013). Cyklisthatet är livsfarligt [Cycling hate is fatal]. *Expressen*, June 7.

Kronsell, A., Smidfelt Rosqvist, L., & Winslott Hiselius, L. (2015). Achieving climate objectives in transport policy by including women and challenging gender norms: The Swedish case. *International Journal of Sustainable Transportation*. https://doi.org/10.1080/15568318.2015.1129653.

Lee, D. J. (2016). Embodied bicycle commuters in a car world. *Social and Cultural Geography, 17*(3), 401–422.

Lerner, T. (2016). För små ytor i trafiken skapar bråk och irritation [Too small spaces in traffic creates fuss and irritation]. *Dagens Nyheter*, April 12.

Lindkvist Scholten, C., Koglin, T., Hult, H., & Tengheden, N. (2018). Cykelns plats i den kommunala planeringen [The place for bicycles in municipalities planning]. K2 Working papers 2018:5.

Ljones, E. (2015). Bilist attackerade cyklister [Car driver attacked by cyclists]. *NVP.se*, May 15. http://www.nvp.se/Arkiv/Artiklar/2015/05/Bilist-attackerade-cyklister/

Lugo, A. E. (2013). *Body-city-machines: Human infrastructure for bicycling in Los Angeles*. Unpublished PhD dissertation, University of California.

Lugo, A. E. (2016). Decentering whiteness in organized bicycling: Notes from the inside. In A. Golub, M. Hoffmann, A. Lugo, & G. Sandoval (Eds.), *Bicycle justice and urban transformation: Biking for all?* (pp. 180–188). Abingdon: Routledge.

McCarthy, D. (2011). "I'm a normal person": An examination of how utilitarian cyclists in Charleston South Carolina use an insider/outsider framework to make sense of risks. *Urban Studies, 48*(7), 1439–1455.

Mellström, U. (2004). Machines and masculine subjectivity: Technology as an integral part of men's life experiences. In M. Lohan & W. Faulkner (Eds.). Masculinities and technologies. Special Issue. *Men and Masculinities, 6*(4), 368–382.

Mouffe, C. (2005). *On the political*. London: Routledge.

Mouffe, C. (online). Agonistic democracy and radical politics. *Pavilion Magazine*. http://pavilionmagazine.org/chantal-mouffe-agonistic-democracy-and-radical-politics/

Nixon, D. (2014). Speeding capsules of alienation? Social (dis)connections amongst drivers, cyclists and pedestrians in Vancouver, BC. *Geoforum, 54*, 91–102.

Pirttisalo, J. (2015). Vikande trend kan fälla Stockholms cykelmål [Slowdown may put Stockholm cycling goal on hold]. *Svd*, July 25.

Polk, M. (2009). Gendering climate change through the transport sector. *Women, Gender and Research, 18*(3–4), 73–82.

Prop 2008/09:93. Mål för framtidens resor och transporter [Goals for future travel and transport]. www.regeringen.se

Redshaw, S. (2008). *In the company of cars: Driving as a social and cultural practice*. Aldershot: Ashgate.

Salazar, N. (2014). Anthropology. In P. Adey, D. Bisell, M. Sheller, K. Hannam, P. Merriman, & M. Sheller (Eds.), *The Routledge handbook of mobilities* (pp. 55–63). New York: Routledge.

SCB, Statistiska centralbyrån. (2016). *På tal om kvinnor och män; lathund in jämställdhet 2016* [*Women and men in Sweden, facts and figures 2016*]. http://www.scb.se/Statistik/_Publikationer/LE0201_2015B16_BR_X10BR1601.pdf

SOU. (2012:70). *Ökad och säkrare cycling – en översyn av regler ur ett cykelperspektiv* [Increased and safer cycling: An overview of rules from a cyclist perspective]. Stockholm: Fritzes.

Spinney, J. (2010). Improvising rhythms: Re-reading urban time and space through everyday practices of cycling. In T. Edensor (Ed.), *Geographies of rhythm: Nature, place, mobilities and bodies* (pp. 113–127). Burlington: Ashgate.

Steinbach, R., Green, J., Datta, J., & Edwards, P. (2011). Cycling and the city: A case study of how gendered, ethnic and class identities can shape healthy transport choices. *Social Science & Medicine, 72*(7), 1123–1130. https://doi.org/10.1016/j.socscimed.2011.01.033.

Stigel, R. (2015). Cyklister i masskrock efter att ha körts på av smitare [Cyclists in mass crash after being hit by getaway man]. *LS Södertälje*, May 4.

Svedberg, W. (2013). *Ett (o)jämställt transportsystem i gränslandet mellan politik och rätt: en genusrättsvetenskaplig studie av rättslig styrning för jämställdhet inom vissa samhällsområden* [A gender (un)equal transport system in the borderland between Policy and Law – A gender legal study of legal governance for gender equality in certain areas of society]. Dissertation, University of Gothenburg.

Ternebom, E. (2013). Bilister har dålig förståelse för oss cyklister' [Car drivers' lack understanding for us cyclists]. *Borås Tidning*, August 10.

Trafikanalys. (2015). RVU Sverige 2011–2014 – Den nationella resvaneundersökningen [RVU Sweden 2011–2014 – The national travel survey]. Statistik 2015:10.

Trafikverket. (2014). *Safer cycling: A common strategy for the period 2014–2020*. Borlänge: Trafikverket.

Trivector. (2014). *Cykeln och cyklisten – omvärld och framtid* [The bicycle and the cyclist: The surrounding world and the future], rapport 2014:103, Stockholm: Trivector.

Willard, F. E. (1997[1895]). *A wheel within a wheel: How I learned to ride a bicycle; with some reflections by the way*. Bedford: Applewood Books.

CHAPTER 10

Hypermobile, Sustainable or Safe? Imagined Childhoods in the Neo-liberal Transport System

Tanja Joelsson

INTRODUCTION: CHILDREN AND THE TRANSPORT SYSTEM

Children's and young people's travel affects the whole family. Cycling to school and leisure activities reduces the whole family's car use. (Information brochure on Uppsala City's work towards increased cycling)

In many European countries, and especially in densely urbanized areas, an increasing awareness of the detrimental consequences of a motorized transport system has developed. Many governments are now addressing the issue, not only from the perspective of traffic safety but also due to pressing concerns around sustainability or public health (UN 2015). The use of cars for household activities is perceived as one of many problem areas where people's behaviours are targeted, such as in the quote above. In addition, children and young people are highlighted as being particularly important for

T. Joelsson (✉)
Department of Education, Uppsala University, Uppsala, Sweden
e-mail: tanja.joelsson@edu.uu.se

© The Author(s) 2019   221
C. L. Scholten, T. Joelsson (eds.), *Integrating Gender into Transport Planning*, https://doi.org/10.1007/978-3-030-05042-9_10

shifting behaviour within the family. Local transportation politics for children and young people is thus positioned within a family context, and in relation to the route to school or to leisure activities, although the connections with wider societal concerns, such as sustainability, public health and equality, have increased in significance. How policies are formulated is also extremely important for planning, since planners transform policies into practice.

Nearly two million, or around 20% of Sweden's nine-million population, are children (Statistics Sweden 2013), and 87% of the entire population lives in densely built-up urban areas (Statistics Sweden 2017). Children are the main users of the local environment close to home, and most of their movement occurs to and from school, to leisure activities and friends, or to different places in the immediate neighbourhood. Given the local character of children's movement, regional and municipal politics becomes particularly important in relation to children's transport and mobility-related issues. Research has identified a decline in children's movement in many Western countries, interpreted as the effects of socio-cultural changes in parallel with the spatial effects of an 'automobility regime' (Böhm et al. 2006; cf. Björklid and Gummesson 2013; Hillman et al. 1990; Fyhri et al. 2011; Kyttä et al. 2015). Research on children's mobility assumes that independent mobility is important for children's development, health and well-being. According to the Swedish government's transport policy objectives of 2008–2009 (SOU 2008/09: 93), particular importance should be paid to children and young people, with a strong focus on safety and security, and independent mobility. The evaluation by government agency Transport Analysis of the Swedish transport policy objectives of 2008–2009, and its overall assessment of the functional objective regarding children and young people, finds that the work aiming to improve children's mobility, traffic environment and use of the transport system has by and large remained unchanged since 2009 (Trafikanalys 2015). This is by no means only a question for municipal, regional or even national authorities but is an issue relating to children's living conditions and rights on a global scale (cf. UN 1989).

The transport system comprises many interconnected levels: from national, regional and local politics, plans and policies to the execution of these within the local landscape. In this chapter, I will direct attention to regional and local policies on transport because such plans and policies form and inform the local infrastructure and built environment. My particular interest is in whether and how children are addressed in the regional plans and policies for transport infrastructure in Uppsala County, Sweden.

My point of departure for the analysis is children's mobilities and access to the transport system. Hence, I use Uppsala County, in Sweden, as a case study in order to illustrate whether and how children are constructed as subjects in policies related to transport infrastructure and municipal policies on cycling and public transport. The questions I have asked in the analysis are what rationales can be identified in the policy documents, and how are children conceptualized within them? What kind of political demands can these rationales enable and, consequently, what is left out? And finally, how is childhood imagined and constructed in regional and municipal transport policy? In the concluding section, I relate policy-making in transportation politics to constructions of childhood in general, and to children's mobility in particular. Moreover, the chapter's findings are discussed in connection with how transport planning can better integrate children's need to move safely.

Children in Transport Policy and in Research on Transport and Mobility

In research with children, mobility has recently emerged as an important theme to investigate (Barker et al. 2009). Mobility, defined as geographical movement or as travel/transport, has been addressed in research with and on children in diverse ways, but mostly in passing and seldom as the main focus of investigation. However, I have found little research dealing particularly with children in transport policy, either in the Nordic countries or internationally. The works referred to below will therefore mostly deal with how children or childhoods have been investigated in both quantitative and qualitative transport and mobility research.

In Sweden, the tendency to highlight traffic safety permeates much of the existing research on children's travel and transport. Although sparse, some quantitative, survey-based research on children and transport in Sweden does exist, focusing on such issues as travel habits (Trivector 2007) or the route to school (Trafikverket 2012). In Sweden, national studies on travel habits are carried out annually or biannually, and children from the age of six are included, but interviewed through a parent. In these studies, children or their spokesperson[1] report that the children make on average 3.1 trips a day, mostly by bike (34%) or by car (33%). The results show that the respondents' results differ depending on the size of the city, the neighbourhood they live in, type of housing, age, trip length and access to a car. Moreover, almost 70% of the children in urban areas who live close to the city centre travel by foot and by bike, whereas the equivalent is 40% for children in rural areas, where motorized mobility (car or bus) is by far the most prevalent. In comparison, 95% of children, regardless of place of residence or size of city, had access to a bike. In the overall study population, 29% of the children make their trips unaccompanied and 28% with their sibling or a peer, compared to 43% who are adult accompanied (Trivector 2007).

The scarcity of quantitative studies on children's travel habits and patterns, both in the Nordic countries and internationally, is problematic in many ways, not least due to the topic of this chapter, resulting in a lack of detailed knowledge at an aggregate level about how, why and where children move (cf. Näsman 1994; see also Qvortrup 2000). Making comparisons over time and across space—as well as studying the impact of socio-economic factors, or of gender, ethnicity and race, (dis)ability and so on—for children as a group, would be an invaluable contribution to the wider knowledge about children's mobility.

Critical research on children and transport policy was sparked by the seminal work of Hillman et al. (1990, cf. Hillman 1993) during the 1990s in England and (West) Germany, replicating survey studies from the 1970s in England, where children's quality of life is understood through the lens of their mobility. Children's mobility is contrasted with that of adults, with reference to personal independence and autonomy, and is also considered in relation to the survey results from 1970s' England and then contemporary Germany. Hillman (1993: 16) concludes:

> The survey findings point to a serious impact, largely attributable to the growth of traffic, on the quality of life of children in the UK. They are increasingly being denied a basic right – to get around on their own – that their parents, and even more their grandparents, enjoyed when they were children. Previous generations of children spent much less of their time under adult surveillance. Playing in the street and getting about in their local neighbourhood – a traditional locus for children's social and recreational activities and experiences – are more often than not forbidden to them. Allied to this, the restrictions have required parents to spend a steadily rising amount of time escorting and 'minding' their children.

However, these studies seldom address children's own views, perspectives or experiences of their travel or mobility. Another stream of research has therefore emerged in line with Hillman's (1993) work, shedding light on children's mobility from a more qualitative vantage point and focusing on what has been termed children's independent mobility. Studies of children's independent mobility have dealt extensively with children, travel and movement. The primary focus has been on the negative impact of motorized road mobility (and the physical environment built around motorized mobility) on children's opportunities to access and use public space in the Global North. Some childhood researchers have nonetheless questioned the 'independence' of children's mobility (Mikkelsen and Christensen 2009), as well as the idea that their diminished range of mobility necessarily entails a worse quality of mobility. Another layer of critique has addressed the rural-urban divide and pointed to how studies of children's mobility have tended to focus upon urban, middle-class households. This leaves unaddressed questions around unequal access to the transport system, which might be linked to other social and geographical inequalities (cf. Lucas 2012). In addition, the US studies show a higher rate of walking and cycling among low-income and minority youth (McDonald 2008), which indicates that inequalities related to race/ethnicity also affect and are affected by mobility. How this relationship is played out has not been investigated at all in Swedish transportation research, but can indirectly be discerned through research investigating segregation (with a focus on housing and dwelling; see for instance Andersson et al. 2016), and child poverty (Rädda Barnen 2013).

An adjacent field of interest in relation to children's mobility is studies on the daily mobility of households, broken down into different factors contributing to and conditioning the daily mobility of members of the household. Here, dwelling must be understood in parallel with access to finances, the transport system, the labour market, the school system and the school market, access to goods and services, as well as cultural conceptions of leisure, the urban-rural divide and ideal childhoods. Parents' choice of where to live in Sweden is related to norms around the ideal places for children to live and grow up, and factors that are deemed important concern safety and security, and closeness to nature (Stenbacka 2011; Tillberg 2001; Sandberg 2012). Access to local school markets due to the free school choice (Gustafson 2011) and cultivating leisure activities (cf. Laureau 2003; Forsberg 2009) are also factors taken into account by parents. For some time now, a trend in which families prefer to live in cities rather than move to the countryside has been identified (Boterman et al. 2010; Carroll et al. 2011). Inner-city parents prefer what Sandberg (2012) has termed a city-social outdoor life, in which proximity to a range of cultural activities is as important as being in the vicinity of children's play areas. Another ideal identified by Sandberg (2012) is the urban-ecological outdoor perspective, by which inner-city parents are keen to offer their children opportunities for play and interaction with 'nature' (here understood as green areas such as woods, forests, meadows, lakes etc.). One important reason for these parents to live in the inner cities was the possibility of being mobile without a car.

In line with a lot of childhood research, research on childhood and mobility has taken two directions. One coincides with the 'parochial locus' identified by Ansell (2008): mobility within the immediate neighbourhood, the journey to school or to sites aimed at children (compare research on transport or traffic safety discussed at the beginning of the section). Attention has been directed towards the neighbourhood, the dwelling or the locality, rather than mobility or movement. The other direction has concerned children and migration and transnational movement (Skelton 2009), and although I do recognize the links and conditional differences between children in various parts of the world, I have not been able to further comment on the subject in this chapter.

In conclusion, children have been sparsely investigated within conventional transport research, but in childhood research mobility has received increasing attention and in research on mobility children's worlds have begun to attract interest (cf. Barker et al. 2009). My contribution with this chapter has the aim of bringing the fields of conventional transport research together with studies on social and cultural geographies of mobility and the geographies of childhood in order to improve the possibilities for more informed decision-making and thus planning.

UNDERSTANDING CHILDHOOD, CHILDREN AND MOBILITY IN THE LIGHT OF URBAN CITIZENSHIP AND CIVIC LEARNING

'Childhood' is a 'modern' conceptualization of the phase between birth and adulthood (Holloway and Valentine 2000). Perceiving childhood as socially and culturally specific is an important foundation for critical childhood research, as is recognizing the historical trajectories underpinning contemporary understandings of children and childhood. Qvortrup (2000: 73) refers to two concepts drawn from the work of German childhood researchers Zinnecker (1990), Zeiher (2003) and Zeiher and Zeiher (1994), which he claims can shed light on how modernity has influenced ideas of childhood:

namely domestication (Verhäuslichung) and insularization (Verinselung). Both these concepts are related to children's space: domestication refers to the observation that, historically, there has been a general trend towards having children removed from streets and other open areas and their being confined to limited spaces protected by fences, walls, etc. (see Zinnecker 1990). Institutionalization is only a special case of this trend. Insularization makes reference to another secular trend, namely that children's open and greater mobility have been replaced by 'islands' in different parts of the city due to a growing differentiation of functions. (see Zeiher and Zeiher 1994: 17ff)

The concepts of domestication and insularization provide a framework for how to both understand and analyse children's spaces and childhood in many contemporary societies in the West. Analytically, then, childhood as a concept is not to be conflated with children, if childhood is used as a concept to denote a macro-perspective on 'the commonality of the social, cultural and economic circumstances that derive from children's minority status and living conditions' (Christensen and James 2000: 4). Qvortrup (2000) further argues that macro-comparative perspectives on childhood are essential for situating qualitative and micro-oriented studies on children's everyday lives. Another key epistemological, methodological and theoretical point in this chapter is therefore that children are competent social actors who are aware of and can influence their own lives (James and Prout 1997; James and James 2004), and as social actors they are therefore involved in a process of creative appropriation and are able to interpret and reproduce society (Corsaro 1992). Involving children in policy-making or spatial planning, and indeed research, concerning children's lives, is one way of shifting the epistemological paradigm whereby the partiality of perspectives and situatedness of knowledge production is laid bare (Haraway 1990). This does not mean, however, that children's voices can be or are heard in all the different arenas and spaces which profoundly shape their lives, pointing to the importance not only of including children's experiences but also of investigating the processes and flows that impinge upon and shape children's lives at a distance (Ansell 2008). The theoretical focuses on how mobility is co-constituted with age is, therefore, a recent and most welcome addition to childhood geographies and research on mobility (Barker et al. 2009; cf. Skelton 2009).

In this chapter, mobility is defined as movement in space, in line with one of Urry's (2004b: 28) five definitions of interdependent mobilities: 'corporeal travel of people to work, leisure, family life, pleasure, migration, and escape'.[2] As noted in the introduction to this anthology, the legacy of feminist research has been and is important in transforming research and policy within transport and mobility, as well as related areas such as spatial planning and land use. According to Law (1999), the field of transport geography has remained largely untouched by discussions on diversity and difference forged by advances in social theory. Conventional transport research is often founded on 'econometric models of traveller behaviour based on methodological individualism and a view of transport as if it is just an exchange-based transaction' (Root 2003: 1; cf. Urry 2004a) and hence rather instrumental in character. It is often also based within the applied sciences, aiming to provide new technologies and systems without questioning ontological assumptions. Research on mobility, on the other hand, concerns the study of movement from a socio-spatial and sociocultural perspective, taking an interest in practices of movement and mobility, and the meaning-making processes related to them. According to Law (1999: 568), this entails transport being given a 'new framing [...] within social and cultural geographies of mobility'. I use the concepts of transport and mobility in parallel, and they should be understood within the framework of the social and cultural geographies of mobility (Law 1999; Introduction, this book).

Mobility is, nevertheless, an ambivalent concept, with the potential for mobility and ability to be mobile making up two sides of the same coin. Mobility simultaneously creates possibilities even as it limits actions (Freudendal-Pedersen 2009). I align my research with traditions of study on children's mobilities which emphasize the cultural, social and spatial constraints that condition and regulate children's movement in public space, and show how children themselves interpret, practise and transform their mobility (see for instance Barker et al. 2009; Mikkelsen and Christensen 2009; cf. Holloway and Valentine 2000; Valentine 2004).[3] In addition, practices relating to transport and mobility are further positioned in this chapter within a framework of the right to the city (RTTC). The notion of RTTC emanates from the idea that mobility is imbued with politics, thus reframing mobility and mobility practices as political acts rather than instrumental or technical exercises (Levy 2013; cf. Law 1999; Mouffe 2013; Bacchi 2009). Mobility, and its related practices, are hence part of what constitutes urban citizenship—'the right to be in and of the city' (Lefebvre 1991: 53). Practising urban citizenship is contingent upon Lefebvre's (1991) idea of both participating in and appropriating space, which in the context of transport and mobility has several implications.

Appropriation, according to Levy (2013: 54), deals with both accessibility and mobility, where transport accessibility is thought of:

> not only in geographical terms as the distance between different locations in the city involving movement at different levels of public space. It also encompasses economic concerns related to affordability, sociocultural aspects related to safety and security in public space, and the physical issues of comfort and ease of design in the use of transport and its related infrastructure.

The right to mobility, on the other hand, is an end value in itself, and understood as ensuring opportunities for everyone to be mobile. The ability to be mobile, however, does not rest upon the idea that the more mobility, the better. The current planning paradigm in most parts of the world has often conceptualized mobility in a certain (often motorized way), thus affecting good mobility for some as being also a sufficient condition for accessibility. When it comes to children, as this chapter will show, sociocultural norms relating to age are fundamentally enmeshed in how transport and mobility are conceptualized in transport policy, thus affecting people's everyday mobility practices.

ANALYSING TRANSPORT POLICY

The material for this policy analysis is *The County Plan for Regional Transport Infrastructure 2014–2025* (Regional Federation of Uppsala County 2014, parts 1 and 2), *The Regional Bicycle Plan for Uppsala County* in Sweden (Regional Federation of Uppsala County 2010), *The Action Plan for the Work with Bicycle Traffic* (Uppsala Municipality 2014a) and *The Action Plan for Public Transport in Uppsala City 2015–2030* (Uppsala Municipality 2014b). I chose county plans because they deal with the question of the overall transport system in Uppsala County and have a mandate to make decisions affecting the region's transport infrastructure. The county plan for regional transport infrastructure sets the agenda for how the transport infrastructure is planned and executed in relation to other, neighbouring counties, and to different objectives at the national and regional level. I included *The Action Plan for the Work with Bicycle Traffic* and *The Action Plan for Public Transport in Uppsala City 2015–2030* because children are referred to walking, cycling and public transport for getting around on their own. Since there is no action plan for walking (cf.

Horton et al. 2014), the action plans regarding public transport and cycling fall into the category of 'child-related' documents. Furthermore, these action plans govern the local municipal work on public transport and cycling: identifying, handling and implementing measures to improve accessibility and the safety of cyclists and users of the public transport system. Other important documents at the municipal level are the *Comprehensive Plan 2010, Policy for Sustainable Development* and works related to public health, but these are not included in the policy analysis.

Uppsala municipality is located in the mid-east of Sweden, with a population of around 215,000 inhabitants, which makes it the fourth largest municipality in Sweden (Statistics Sweden 2016). Uppsala is one of the fastest growing regions in Sweden and has a population density of 98 inhabitants per square kilometre (national average [NA] 24.5; Statistics Sweden 2016). Around 21% of the municipality's population consists of children under 18 years of age (Uppsala Municipality Statistics 2016). Uppsala is the county town of Uppsala County, and hosts the oldest university in the Nordic region (which is the third largest employer in the municipality). Uppsala County borders five other counties, among them, Stockholm County. Compared to the national average, the unemployment rate is lower (4%; NA 6%), and the majority of the population has upper secondary education (35%; NA 43%) or higher education (55%; NA 45%).[4] Like many Swedish cities, the county town is planned for motorized mobility, and struggles with how to transition from this car-based built environment to more sustainable modes of transport (cf. Emanuel 2012).

The methodological framework for the policy analyses is inspired by Bacchi's (2009) discursive approach to policy problems: what is problematized in national, regional and municipal policies around transport? Bacchi (2009) argues that the *problematizations,* in relation to policy, define and shape what we see as 'social problems'. The 'issues that are problematized – how they are thought of as 'problems' – are central to governing processes' (Bacchi 2009: xi). The focus is therefore on the underlying assumptions relating to children, which often become visible in discursive tensions, conflicts, breaches or boundary-crossings—or silences and absences—in the policy material. Bacchi (2009) is inspired by Foucault's (1990) genealogical method, where the interest lies in how discourses are produced—how certain phenomena are talked about and addressed and, equivalently, what is left out of that talk. Discourses shape the ways in which phenomena are talked about and construct a framework for what can be expressed about the topic and how subjects (and objects) are constituted (Bacchi 2009).

Discourse is not reduced to language, but exists outside language and can therefore fruitfully be seen as *a framework for thought* (Westerstrand 2008) or *a figure of thought* (Lundgren 2004). Lundgren (2004: 93, emphasis in original, my translation) contends that it is:

> *the figure of thought* [...] that is important to grasp when talking about discourse, more particularly the underlying, and especially the unspoken and taken-for-granted, figure of thought.

Competing, and sometimes conflicting and contradictory, discourses can be sorted into discourse orders (Fairclough 2003). The discourse order of children's mobility can hence be understood to encompass several different ways of interpreting, understanding and talking about children's mobility. As I will discuss in the policy analysis, within the policy field, parallel or altogether different discourses, compared to those current in society at large, can be identified. One way to grasp how a discourse is constituted is to identify when, where and how boundaries are breached. When norms are breached, a reaction is often invoked, and these reactions reveal the cultural conceptions and normative framings that give the discourse its contours.

For clarity, when I talk about the general underlying logics permeating policies, I refer to them as *rationales*. In my view, there is no theoretical distinction between rationale and discourse, but for the sake of the argument (and for linguistic purposes), I have found it necessary to make an analytical distinction in this chapter. The general logics are then referred to as rationales, whereas the particular logics concerning children are referred to as discourses.

RATIONALES IN TRANSPORT POLICY

The rationales that permeate the policies can be grouped into an economic rationale, foregrounding the hypermobile subject; a caring rationale, foregrounding a sustainable subject; and a risk rationale, foregrounding a safe subject. The first rationale is most clearly articulated in the county plans, whereas the second is located at the municipal level. The third rationale—that of risk—is addressed in both the regional and municipal policies, but in somewhat different ways. Furthermore, two discourses relating to children's mobility in transport policy within these rationales are taken up: children as absent and children as particularly worthy of protection.

The Economic Rationale: Economic
Growth as the Guiding Principle

The chairperson of the public transport committee in Uppsala recently argued in a letter to the editor that 'the city is a hub for (economic) growth, which demands improvements in the bus and train services' (Uppsalatidningen 2016). This quote is illustrative, and points to the need for macro-analyses of transport politics. In order to understand the regional policies for infrastructure, a macro-oriented perspective is needed. Since the early 2000s, the expansion of labour-market regions has been an important objective in Swedish regional policy, built on the idea that increased commuting contributes to regional development and to 'the attractive region' (Amcoff 2007). Shifts in the labour market towards what has been termed a service economy give rise to new challenges, mostly affecting sparsely built-up areas (Illeris 1996). Due to systematic planning within a functionalist paradigm, (larger) cities cater for the bulk of employment opportunities and the outskirts become residential areas, with car dependency as one consequence. In Koglin's (2017) words, the 'materialities for mobilities' are encouraging the use of certain (often motorized) modes of transport, while the use of alternatives are discouraged due to infrastructure and the built environment, or a lack of developed transport services. That being said, the expansion of labour-market regions affects the spatial relations within the regional geography, contributing to geographical inequalities. One effect of the politics associated with stimulating economic growth in the regions is not only that some people or groups of people are expected to commute longer distances from sparsely built-up areas to the urban centre, but also that this is framed partly as self-evident, partly as a necessity in order to meet the needs of the market.

The conditions for planning are hence linked to the idea of the expansion of labour-market regions in Sweden. This expansion has been formulated in conjunction with regional economic growth, which can be measured through commuting to work (SOU 2004: 34), but can also be seen in relation to the EU's innovation policies, which have the aim of promoting knowledge, technical development and innovation. The EU's policies, in combination with the Swedish government's aims, contribute to the need many municipalities feel to be 'part of the race', which often means that considerable efforts are made in the quest to attract people (labour force) and trade and industry (investment). The economic rationale can be framed as an overarching discourse within which economic growth is seen as central. Regions are understood and analysed in terms of their contribution to national economic growth, and are primarily regarded as labour markets. The strategic orientation of regional policies is in line with this overall tendency, formulated in the vision: 'Uppsala is Europe's most attractive region of knowledge' (Regional Federation of Uppsala County 2014: 7; cf. Mukhtar-Landgren 2008).

The economic growth of the region has a spillover effect on the transport system, leading to policy goals for the transport infrastructure which focus on both maintaining and increasing its capacity. Two of the seven policy goals in regional policy potentially relate to children: the proportion of child users of the public transport system in the region is to be doubled by 2020; and the number of trips by foot or bicycle in the region is to increase. The sub-goals presented beneath the overall strategic aims are all positioned, in one way or another, within the economic framework. The focus is on increasing the capacity of the rail network and road system in order to meet the needs of increased work commuting. When cycling is discussed, it is in reference to commuting and being able to take a bike on different forms of transport in order to ease multimodality in travel.

The framework of economic growth is hence centred around the needs of the labour market, with its built-in urban bias. Under this rationale, the ideal subject is hypermobile, able to travel across geographically extended areas (cf. Whitelegg 1993). The framework wherein the expansion of labour-market regions is discussed relates to economic growth and the interests of trade and industry, leaving the individual and their everyday life aside (Gil Solá 2013). Hypermobility is not conceptualized here as having a negative impact on personal health, social relations, place affinity or other aspects relating to quality of life, although qualitative studies show far-reaching consequences for the experience of quality of life for long-distance commuters (Jönsson and Lindkvist Scholten 2010; Gil Solá 2013; Sandow, this volume; cf. Adams 1999, 2005). Due to its focus on the labour market, the rationale of economic growth cannot encompass age diversity in a successful manner. The adult bias has the effect of obscuring children (but also other groups of people who do not have a connection with the labour market).

However, given the ideal and goal of the expansion of labour-market regions, which also pertain to geographical radius and distance, it is safe to contend that while these policy goals may relate to children, it is more likely that in reality they do not. In general, children's mobility tends to be more local, meaning that they are moving around (on their own) to a greater extent in their immediate neighbourhood. On the other hand, even though research on the expansion of labour-market regions has not been carried out on or with children (cf. Gil Solá 2013, who discusses research on the effect of the expansion of labour-market regions on women's mobility patterns), some indications can nevertheless be formulated. In some parts of Sweden, the expansion of labour-market regions has had the effect of eroding basic public services in rural and peri-urban communities. This affects children and young people, for instance in relation to access to the transport system (Joelsson 2013), or access to the school system due to the closure of local schools (Cedering 2016). Cedering (2016) discusses how school closures entail an increased car dependency for families living in the countryside, leading to longer school journeys for the affected children. Increased car dependency and longer school journeys are also, however, a more general effect of the implementation of free school choice in Sweden in the 1990s (Boverket 2017), affecting urban as well as rural children, as well as being a consequence of shifts in risk-management practices and parenting cultures in relation to children's everyday mobilities (Joelsson forthcoming b, see also Tillberg 2002). Nevertheless, children are potential users of the county's transport system, although these circumstances are not further mentioned in the policy documents.

Given the economic framework of regional policies, the category of 'children' does not fit neatly into the policies on transport infrastructure. The most pervasive discourse on children in transport policies is that of absent children. In the political discussion on transport, children remain largely *unrecognized as a party*, with interests that might diverge from or conform to those of the adult (working) population. Mobility is hence framed primarily as an adult asset in relation to the labour market, where greater mobility is equated with autonomy and power. Children, in this sense, are political non-subjects. The absence of children and their lack of subject position is not in itself a novel finding and has been discussed and debated from various vantage points within childhood studies, not least in relation to children's rights and children as political subjects/actors. Not being recognized as a political subject has of course also spilled over into areas related to transport, such as spatial planning and land use.

A Caring Rationale: Constructing a Sustainable Subject

The second rationale which can be identified in the policies is concerned not only with providing public transport and municipal infrastructure for the city's population, but also with addressing so-called soft issues in the transport system: gender equality, and social equality in general, accessibility and safety (social sustainability). The policy documents also address issues of ecological sustainability. I have therefore chosen to refer to the rationale, or the discourse, as one of caring, which is to be understood as addressing issues related to the welfare state: the provision of equal services for as many groups of people in the population as possible (see Introduction). This rationale does not primarily target the individual, like the economic discourse above, but tends to meet the needs of groups or collectives. The *Action Plan for the Work with Bicycle Traffic* deals with the municipality's work on cycling. The plan singles out three target groups to be prioritized: the working population, students, and children and young people. The Action Plan further mentions that in order to get children and young people to start cycling, a grid of bicycle lanes and parking possibilities is needed. Targeting children and young people is seen as a way to increase the levels of cycling, increasing comfortability and improving traffic safety. In addition, cycling is also perceived to increase children's learning and promote health. For instance, the plan for bicycle traffic states that:

> Children and school youth are a particularly important group because their travel affects the whole household. If they are given good preconditions for cycling to school and leisure activities, car use can be reduced. [...] In order to meet the conditions [that children and young people should be able to get around easily; stated in the Child and Youth Policy Programme 2010–2014], Uppsala needs to systematize its work with children and young people's travels. The municipality should work actively on safe routes to school and encourage parents of small children to cycle to school and leisure activities. (Uppsala Municipality 2014a)

Uppsala municipality recognizes the need to systematize work with children's travel. However, there is a strong tendency to conceptualize the family unit as a target for specific measures, which is most apparent when children and young people are discussed. Children are conceptualized as becoming adults, making it particularly important to target the family and their habits, in order to ensure a more sustainable adult in the future. The subject who is constituted within the caring rationale is the sustainable subject, who has the potential to shift from unsustainable travel habits to more sustainable modes (public transport, walking, cycling). In comparison with the economic rationale, the individual is here cast in a more holistic light, travelling not only to work, but also to leisure activities, school, shops and to visit friends and family. Behavioural changes, through information and positive reinforcement, are deemed important for shifting towards more sustainable societies, but physical measures to decrease the use of the private car are proposed in parallel.

The Risk Rationale: A Safe Transport System

In order to get more children and school youth to start cycling, a good bicycle network and good parking opportunities are needed. Measures targeting the behaviours of both children/school youth and their parents are needed. *It is primarily safety and security that characterize a good cycling network for this group.* Measures directed at this target group lead to both increased cycling (which in turn has health benefits) and to travel time benefits, increased comfort and increased traffic safety for those who cycle. Cycling is also beneficial for learning through physical activity. (Uppsala Municipality 2014a: 14)

The final rationale I have identified can be characterized as a risk rationale. This rationale is present in both of the former discourses, but less so in the economic rationale and more in the caring rationale. The risk rationale refers to how it is primarily traffic safety that is taken up and presented in the policies, and thus how the safe subject is constituted. Even though the policies for transport infrastructure are firmly positioned within the overall framework of economic growth and regional expansion, children are mentioned in particular instances. Paradoxically, then, the discourse of invisible children can embody another discourse where children are constituted as *particularly protection-worthy*. Children's mobility is here reduced to a matter of traffic safety: children are unprotected road users in general and/or road users with the school as a destination in particular. The route to school is hence constructed as explicitly connected to children's mobility.

The regional policy strives to encapsulate a progressive approach in that it:

suggests a raising of ambition in this county plan when it comes to building up the regional pedestrian and cycle network, especially on the sections where we know that many people walk and cycle. In our selection model, children and their way to school are particularly taken into account. (Regional Federation of Uppsala County 2014: 33)

Children are mentioned in relation to unprotected road users and a discussion on traffic accidents at the end of the County Plan. Shortly thereafter, a whole section is devoted to them, under the heading

'Children'. Although it is mentioned that children's needs probably entail a different design for the transport system as a whole, the plan concludes that this would be too costly. It is not possible for the region to finance a rebuilding of the entire grid of cycle and pedestrian lanes. The authors also point out that taking the children's perspective stands in conflict with other objectives, such as the demand for fast transport of goods and people. It is noteworthy that this is the only section in the whole plan where conflicts between different objectives are explicitly addressed.

Besides the regular ongoing work, which was in particular focus during 2015, one of the five areas for measures addressed in *The Action Plan for the Work with Bicycle Traffic* is safer routes to school. However, this is done within the caring rationale, mentioned above. Nevertheless, the journey to school is an important part of children's daily mobility, due to the fact that school is mandatory in Sweden. A mandatory school system implies that the journey to school is also a mandatory aspect of each school day. Many schools recommend certain routes to travel to school, most often that the children should walk, cycle or use public transport. Because of this, the routes to school should be considered as an extension of school space.

According to the journey-to-school study by Trafikverket (2012), 59% of children travel less than 2 km to their school, but 46% of parents consider their children's route to be unsafe. The number of children who cycle to school has decreased over the years during which the survey has been carried out, with variations depending on the season: 48% of children walk or ride their bicycle to school during the winter season (November–March) and 58% during the summer season (April–October), compared to 24% who are chauffeured (an increase from previous surveys). The conclusions of the report are in line with another study (Trivector 2007), which shows that more children are being accompanied by an adult to or from school and that fewer children walk or cycle to school.

Although the focus on the journey and route to school is important, it is also problematic that one of the few instances where children are addressed in transport policies is in relation to school routes. This is often linked to the ambition to reduce traffic accidents along the stretches where children are imagined to walk or cycle. The association between children and school routes (and in addition to traffic safety) compartmentalizes children, contributing to what has been referred to as their domestication and insularization (Zinnecker 1990; Zeiher and Zeiher 1994). Law (1999) argues that the focus in research on adults' transport and mobility in work travel is not surprising given the status of work travel in society as a vital feature of urbanism: an actual and metaphorical bridge between the public and private, production and reproduction, work and home. Similarly, I would argue, the route to school functions as the paradigmatic emblem of the imagined and actualized connection between children's workplace— the school—and their homes. The specific focus on the route to school can thus be perceived as the only intelligible way of conceptualizing children within the adultist framework of economic growth, regional expansion and transport planning.

Safety is here primarily perceived as traffic safety, although existing research with children shows how they talk about the social dimensions of safety when moving around in public (Joelsson forthcoming a). There is a tendency to disregard power in the realm of traffic, thus never confronting the motorized norm. In relation to children, traffic safety becomes narrowly defined as distance to, or separation from, other, often motorized, modes of transport.

Nevertheless, children are seen as central within this rationale, and are also cast as a group for whom particular measures should be taken. The fact that children are conceptualized as beings instead of becomings—as an important group within the population that needs current protection— is, however, double-edged: children are constituted as particularly protection-worthy, cast as dependent and vulnerable. This manoeuvre does not, however, challenge the structural conditions that in fact produce children as vulnerable, which contributes to a separation of children from the perceived adult public world and sphere.

THE ADULT MOTORIZED NORM IN TRANSPORT POLICY: THE PARADOX OF CHILDREN

My analysis illustrates how the different rationales in transport policy are governed by different interests: economic growth, sustainability and traffic safety. These are in part mutually exclusive, in that the goal of economic growth which permeates regional policies can be seen as regulating the direction that municipal policies can take. The overall regional infrastructure of the transport system is governed by economic values that exclude the formulation of other values connected to sustainability in particular, unless these values are somehow operationalized into economic measurements. Regional policies on transport infrastructure need to take into account values other than the purely economic, if society is to meet the challenge of equality in the transport system.

According to data gathered from an ongoing ethnographic research project, the first two rationales, the economic and the caring, are often not addressed, discussed or deemed important when parents and children talk about their mobility and their access to public space. The discourse that children and their parents relate to most is the risk rationale (Joelsson forthcoming b). I argue that the existing policies, in that they are driven by diverse interests (economic, sustainability), are unable to address children's everyday lives and what children themselves find important in relation to their mobility (Joelsson forthcoming a). The matters that are not encompassed in the policies are left to parents and children themselves to deal with and manage. In some senses, due to the narrow scope of the risk rationale as merely relating to traffic safety, and more crudely to traffic safety on children's school routes, the safe subject is contained in both space and time. I would suggest that traffic (or mobility in all its forms) is part of public space, instead of perceiving public space as consumed by traffic. All other places, which are not perceived to be traffic-related, are left out of the policies, thus contributing to insularizing children's lives and spaces (cf. Zeiher 2003). There is also no differentiation between children as a group, obfuscating the different needs that children have due to geographical location, socio-economic position, gender, ethnicity, race, ability and age.

The commonality between the policies is the unquestioned status of the adult motorized norm. Although all of the plans address the need to shift to more sustainable modes of travel, the tendency is to leave unaddressed the power asymmetries in society at large, and within the transport system in particular. The effect is that transport and mobility in the policies are primarily understood as motorized transport and mobility, even when, exceptionally, children's mobility becomes the focus. Moreover, transport and mobility are not seen as political in the sense that there is an acknowledgment of the different interests that underpin the policies (see 'Introduction'). The imagined users of the transport system by and large conform to normative and hegemonic social, cultural and spatial conceptions of who occupies public space (Valentine 2004).

TRANSFORMING TRANSPORT POLICY: IMPLICATIONS FOR EQUAL TRANSPORT PLANNING

This analysis illustrates what the diverse regional and municipal policies struggle with: how to think of children as citizens (cf. de Visscher 2014) and the need for improved transport planning. One discursive effect of the construction of children as separate from and subordinated to adults is that the measures proposed are demarcated to particular topics, arenas and places. Children's positioning as political subjects is neither straightforward nor unproblematic. The two discourses I have identified, the discourse of absent children and the discourse of children as worthy of protection, exist in parallel when the subject of children and transport policy is highlighted. These discourses also produce particular subjects, where children are cast as either political non-subjects or apolitical subjects.

Näsman (1994) points out that children are considered subjects neither in a political sense, nor in a social/scientific sense. Due to their age, children have not, until recently, been individualized, a step which makes it possible to discuss them as rights-holders or beneficiaries (Näsman 1994). This has also affected the view of children as not competent enough to decide upon or give input on their lives. Näsman (1994: 186) argues that children's position is ambiguous:

[I]n legislation and in statistics, they [children] are viewed as subordinate to their parents and become invisible as individuals; but both as individual citizens and dependents, they are increasingly the target of state distributive systems of rights, support, and services.[5]

Critiques have been made of an unproblematic celebration of children's agency, which runs the risk of not recognizing the limits and restrictions on children's opportunities to act and engage in the traditional political sphere (Kohan et al. 2015; cf. Ansell 2008; Kjørholt 2008). Children are excluded from participation in decision-making arenas through cultural, spatial and social processes—which position them not only as less knowledgeable but also as less competent. Näsman (1994: 87) contends that children's 'principal way of gaining autonomy and status is growing up'. Nevertheless, although children as a group are excluded from the political realm and/or do not exercise political agency in the conventional sense, the need to discuss children as political subjects is important. My study only touches upon the issue in passing, but it is 'critical' in the sense Alanen (2011) talks of as vital for childhood research. Alanen (2011: 150) notes that the criticality of childhood research must, in addition to remaining reflexive about our research practices, entail 'a normative turn', in which we 'specify what constitutes a good, or at least better life for children and for human beings in general'. This endeavour entails what Ansell (2008) notes as a step further from or beyond the immediate focus on children's individual lives, to the regional, national and global impacts and contexts. I address the notion of the political subject with these tensions in mind. Is it possible to discuss children as actors without romanticizing the scope and range of their agency? Widening the notion of what politics is and how it is done opens up possibilities to discuss political practices. Political practices can encompass discursive as well as material practices, practices that take place within 'a child-centered material spatial ontology', as proposed by Ansell (2008: 199; cf. Cele 2013). Agency can here be understood as performative, relational and becoming, in contrast to the liberal idea of a conscious, rational and controlled subject acting wilfully and with a direction. Cele (2013) has argued that a narrow view of what politics and political participation is tends to exclude children, and constructs them as apolitical and voiceless. Children and young people's use of, and social interaction in, the public, 'produce spaces through which the political is performed' (Cele 2013), where everyday life takes centre stage in re-formulating the political. Aspects of everyday life can be regarded as political, but everyday life is also the arena where children can 'perform and communicate their political subjectivities' (Cele 2013: 85). Accordingly, de Visscher (2014) argues that children's presence in public places can in fact be seen as part of 'civic learning' (Gert Biesta 2011): of how children learn to become fellow citizens. Citizenship is in this sense not built on status, but on 'citizenship as a quality of everyday social practices' (de Visscher 2014: 74, based on Lawy and Biesta 2006). However, in order for children to be able to engage in civic learning, public space needs to be considered safe and secure (enough) by children and their parents. Here, planners have a defining role, since the development of age-friendly public space is in part a matter of planning.

The strength of the RTTC framework in this context lies in highlighting the *political character* of transport, and its potential to cut across differences (Levy 2013).[6] In its most fundamental sense, the RTTC framework in my study gears the chapter's research questions towards issues of children's access to and presence in both the transport system and public space, thus working to deconstruct the tendency to separate traffic from public space, as well as place, mobility and social relations. The framework is also helpful for approaching the question of children's everyday mobilities as a children's rights issue, for every child's right to access public space and to move around in our common environment without feeling restricted, afraid or unsafe. Here, the importance of planning is obvious, as planners engage in weighing different interests against each other, which ultimately decides the future design of the transport system and the built environment (Fainstein and Servon 2005; Rosenbloom 2005).

NOTES

1. In Sweden, research participants under 15 years of age are accessed through their guardians (SFS 2003: 460). In Trivector's report, it is not clear whether the survey was answered by the children or their parents/an adult, although the survey was directed at children in the age range of 6–15 years. The survey was sent out to 7870 children, and the response rate was 54%. The survey by Trafikverket (2012) has been carried out every third year since 2000. The survey is addressed to parents, and in the latest study 2800 children between 6 and 15 years were targeted, with a response rate of 62.1%.
2. Urry's (2004b: 28) taxonomy of five interdependent mobilities includes: 'the corporeal travel of people to work, leisure, family life, pleasure, migration and escape, organized in terms of contrasting time-space patterns ranging from daily commuting to once-in-a-lifetime exile; the physical movement of objects include food and water producers, consumers and retailers; as well as the sending and receiving of presents and souvenirs; the imaginative travel effected through the images of places and peoples appearing on and moving across multiple print and visual media and which then construct and reconstruct visions of place, travel and consumption; virtual travel often in real time transcending geographical and social distance and forming and reforming multiple communities at a distance; communicative travel through person-to-person messages via personal messages, postcards, texts, letters, telegraph, telephone, fax and mobile'.
3. This policy analysis is part of a larger ethnographic research project, in which 59 children, aged 7–13 years, and 33 parents/guardians have been participating. Findings from the ethnographic part of the project are discussed in two other articles: one concerning children's perspectives on their mobilities (Joelsson forthcoming a), and the other addressing parents' risk-management practices and parenting cultures (Joelsson forthcoming b).
4. Commuting to Stockholm County makes up a significant proportion of the labour-force mobility (15,800 individuals commuting from Uppsala compared to 6229 commuting from Stockholm County to Uppsala). In comparison, 10,573 individuals commute to other municipalities within Uppsala County, whereas 3464 individuals from Uppsala commute to other municipalities within the county. See https://www.uppsala.se/contentassets/f09f9e6b994f41408c66064a2da8470b/statistisk-folder_sv.pdf

5. However, the geopolitical situation in Europe and the rest of the world, forcing large groups of people to move (around), challenge and complicate the notion of children as rights-holders and citizens (cf. Kjørholt 2008). Kjørholt (2008: 33) argues that a difference-centred approach to citizenship 'broadens the concept of participation, relating the practice of participation rights to belonging and community, [...] [thus] recognizing children as citizens in a manner that includes their vulnerability and dependency in the concept of citizenship'.

Although the focus of my study is on well-off children with Swedish citizenship, global processes and events influence the national arena. They relate to the ongoing dialogue around protection vs. participation, and the relation to neo-liberalism (see, e.g. Kohan et al. 2015).

6. The RTTC framework has been subjected to critique, notably for its assumed urban and middle-class bias. The RTTC framework in its conventional use can be seen as neglecting questions of age and generation as well, although the theoretical space for such elaborations is easily available.

References

Adams, J. (1999). *The social implications of hypermobility.* Paris: OECD.

Adams, J. (2005). Hypermobility: A challenge to governance. In C. Lyall & J. Tait (Eds.), *New modes of governance: Developing an integrated policy approach to science, technology, risk and the environment* (pp. 123–138). Aldershot: Ashgate.

Alanen, L. (2011). Critical childhood studies? *Childhood, 18*(2), 147–150.

Amcoff, J. (2007). *Regionförstoring – idé, mätproblem och framtidsutsikter* [Regional expansion – Idea, problems of measuring and future prospects]. Institutet för framtidsstudier 2007: 7, Retrieved August 7, 2017, from http://www.iffs.se/media/1195/20070620162424filGiCdC2OiL18o0lm9j11M.pdf

Andersson, R., et al. (Eds.). (2016). *Mångfaldens dilemma. Boendesegregation och områdespolitik* [The dilemma of diversity: Housing segregation and neighbourhood politics]. Malmö: Gleerups.

Ansell, N. (2008). Childhood and the politics of scale: Descaling children's geographies? *Progress in Human Geography, 33*(2), 190–209.

Bacchi, C. (2009). *Analysing policy: What's the problem represented to be?* Frenchs Forest: Pearson.

Barker, J., Kraftl, P., Horton, J., & Tucker, F. (2009). The road less travelled: New directions in children's and young people's mobility. *Mobilities, 48*(1), 1–10.

Biesta, G. (2011). *Learning democracy in school and society: Education, lifelong learning and the politics of citizenship.* Rotterdam: Sense Publishers.

Björklid, P., & Gummesson, M. (2013). *Children's independent mobility in Sweden.* Stockholm: The Swedish Transport Administration.

Böhm, S., Jones, C., Land, C., & Paterson, M. (Eds.). (2006). *Against automobility*. Malden: Blackwell Publishing.

Boterman, W. R., Karsten, L., & Musterd, S. (2010). Gentrifiers settling down? Patterns and trends of residential location of middle-class families in Amsterdam. *Housing Studies, 25*(5), 693–714.

Boverket [National Board of Housing, Building and Planning]. (2017). *Skolans nya plats i staden. Kommuners anpassning till skolvalet och urbana stadsbyggnadsprinciper* [The new place of the school in the city: How municipalities adjust to the school choice and to principles for urban city building]. Report 2017:16.

Carroll, P., Witten, K., & Kaerns, R. (2011). Housing intensification in Auckland, New Zealand: Implications for children and families. *Housing Studies, 26*(3), 353–367.

Cedering, M. (2016). Konsekvenser av skolnedläggningar. *En studie av barns och barnfamiljers vardagsliv i samband med skolnedläggningar i Ydre kommun* [Consequences of school closures: A study of children's and families' everyday life in relation to school closures in Ydre municipality]. PhD dissertation, Uppsala University.

Cele, S. (2013). Performing the political through public space: Teenage girls' everyday use of a city park. *Space and Polity, 17*(1), 74–87.

Christensen, P., & James, A. (2000). Introduction: Researching children and childhood cultures of communication. In P. Christensen & A. James (Eds.), *Research with children: Perspectives and practices* (pp. 1–9). Abingdon/New York: Routledge.

Corsaro, W. (1992). Interpretive reproduction in children's peer cultures. *Social Psychology Quarterly, 55*(2), 160–177.

de Visscher, S. (2014). Mapping children's presence in the neighbourhood. In G. Biesta et al. (Eds.), *Civic learning, democratic citizenship and the public sphere* (pp. 73–89). Dordrecht: Springer Science.

Emanuel, M. (2012). *Trafikslag på undantag. Cykeltrafiken i Stockholm 1930–1980* [Transport mode excepted: Bicycle traffic in Stockholm 1930–1980]. PhD dissertation, KTH Royal Institute of Technology.

Fainstein, S. S., & Servon, L. J. (2005). *Gender and planning: A reader*. New Brunswick: Rutgers University Press.

Fairclough, N. (2003). *Analysing discourse: Textual analysis for social research*. New York: Routledge.

Forsberg, L. (2009). *Involved parenthood: Everyday lives of Swedish middle-class families*. PhD dissertation, Linköping University.

Foucault, M. (1990). *The history of sexuality. Volume 1: An introduction*. Harmondsworth: Penguin.

Freudendal-Pedersen, M. (2009). *Mobility in daily life: Between freedom and unfreedom*. London/New York: Routledge.

Fyhri, A., et al. (2011). Children's active travel and independent mobility in four countries: Development, social contributing trends and measures. *Transport Policy, 18*, 703–710.

Gil Solá, A. (2013). *På väg mot jämställda arbetsresor. Vardagens mobilitet i förändring och förhandling* [Towards gender equality? Women's and men's commuting under transformation and negotiation]. PhD dissertation, University of Gothenburg, Sweden.

Gustafson, K. (2011). No-go area, no-go school: Community discourses, local school market and children's identity work. *Children's Geographies, 9*(2), 185–203.

Haraway, D. (1990). *Simians, cyborgs, and women: The reinvention of nature*. London: Free Association Books.

Hillman, M. (1993). One false move... An overview of the findings and issues they raise. In M. Hillman (Ed.), *Children, transport and the quality of life* (pp. 7–18). London: Policy Studies Institute.

Hillman, M., Adams, J., & Whitelegg, J. (1990). *One false move... A study of children's independent mobility*. London: Policy Studies Institute.

Holloway, S. L., & Valentine, G. (2000). *Children's geographies: Playing, living, learning*. London/New York: Routledge.

Horton, J., et al. (2014). 'Walking... Just walking': How children and young people's everyday pedestrian practices matter. *Social & Cultural Geography, 15*(1), 94–115.

Illeris, S. (1996). *The service economy: A geographical approach*. Chichester: Wiley.

James, A., & James, A. L. (2004). *Constructing childhood: Theory, policy and social practice*. Hampshire/New York: Palgrave Macmillan.

James, A., & Prout, A. (Eds.). (1997). *Constructing and reconstructing childhood*. London: Falmer Press.

Joelsson, T. (2013). *Space and sensibility: Young men's risk-taking practices with motor vehicles*. Dissertation No. 574, Linköping University, Linköping.

Joelsson, T. (forthcoming a). "I get a whiz in my body as I walk past it": Children's views of their everyday spatial mobilities. Conditionally accepted to *Mobilities*.

Joelsson, T. (forthcoming b). "So we don't spoil them": Understanding children's everyday mobility through parents' affective practices. Conditionally accepted to *Children's Geographies*.

Jönsson, S., & Lindkvist Scholten, C. (2010). *Påbjuden valfrihet? om långpendlares och arbetsgivares förhållningssätt till regionförstoringens effekter* [Mandatory freedom of choice? Long-distance commuters' and employers' approach to the effects of regional expansion]. Linnéuniversitetet, Institutionen för samhällsvetenskaper.

Kjørholt, A. T. (2008). Children as new citizens: In the best interest of the child? In A. James & A. L. James (Eds.), *European childhoods: Cultures, politics and childhoods in Europe* (pp. 14–37). Basingstoke/New York: Palgrave Macmillan.

Koglin, T. (2017). Urban mobilities and materialities: A critical reflection of "sustainable" urban development. *Applied Mobilities, 2*(1), 32–49.

Kohan, W. O., Olsson, L. M., & Aitken, S. (2015). "Throwntogetherness": A travelling conversation on the politics of childhood, education and what a teacher does. *Revista Electrônica de Educação, 9*(3), 395–410.

Kyttä, M., et al. (2015). The last free-range children? Children's mobility in Finland in the 1990s and 2010s. *Journal of Transport Geography, 47*, 1–12.

Laureau, A. (2003). *Unequal childhoods: Class, race, and family life.* Berkeley: University of California Press.

Law, R. (1999). Beyond "women and transport": Towards new geographies of gender and daily mobility. *Progress in Human Geography, 23*(4), 567–588.

Lawy, R., & Biesta, G. (2006). Citizenship-as-practice: The educational implication of an inclusive and relational understanding of citizenship. *British Journal of Educational Studies, 54*(1), 34–50.

Lefebvre, H. (1991). *The production of space.* Oxford: Basil Blackwell.

Levy, C. (2013). Travel choice reframed: "Deep distribution" and gender in urban transport. *Environment & Urbanization, 25*(1), 47–63.

Lucas, K. (2012). Transport and social exclusion: Where are we now? *Transport Policy, 20*, 105–113.

Lundgren, E. (2004). *Våldets normaliseringsprocess. Tre parter, tre strategier* [The normalization process of violence: Three parties, three strategies]. Stockholm: ROKS.

McDonald, N. (2008). Critical factors for active transportation to school among low-income and minority students. *American Journal of Preventive Medicine, 34*(4), 341–344.

Mikkelsen, M. R., & Christensen, P. (2009). Is children's mobility really independent? A study of children's mobility combining ethnography and GPS/mobile phone technologies. *Mobilities, 4*(1), 37–58.

Mouffe, C. (2013). *Agonistics: Thinking the world politically.* New York: Verso Books.

Mukhtar-Landgren, D. (2008). City-marketing in a dual city: Discourses of progress and problems in post-industrial Malmö. In B. Petersson & K. Tyler (Eds.), *Majority cultures and the politics of ethnic difference* (pp. 55–74). Houndmills/Basingstoke/Hampshire/New York: Palgrave Macmillan.

Näsman, E. (1994). Individualization and institutionalization of childhood in today's Europe. In J. Qvortrup, M. Brady, G. Sgritta, & H. Wintersberger (Eds.), *Childhood matters: Social theory, practice and politics* (pp. 165–188). Aldershot: Ashgate.

Qvortrup, J. (2000). Macroanalysis of childhood. In P. Christensen & A. James (Eds.), *Research with children: Perspectives and practices* (pp. 66–86). Abingdon/New York: Routledge.

Rädda Barnen [Save the Children]. (2013). *Barnfattigdom i Sverige* [Child poverty in Sweden].

Root, A. (Ed.). (2003). *Delivering sustainable transport: A social science perspective*. Amsterdam: Pergamon.

Rosenbloom, S. (2005). Women's travel issues. In *Gender and planning: A reader* New Brunswick: Rutgers University Press (pp. 235–255).

Sandberg, M. (2012). *De är inte ute så mycket Den bostadsnära naturkontaktens betydelse och utrymme i storstadsbarns vardagsliv [They are not outdoors that much. Nature close to home – Its meaning and place in the everyday lives of urban children]*. Phd diss, University of Gothenburg.

SFS 2003:460 Lag om etikprövning av forskning som avser människor [Law on ethical review concerning research on humans].

Skelton, T. (2009). Children's geographies/geographies of children: Play, work, mobilities and migration. *Geography Compass, 3*(4), 1430–1448.

SOU. (2004:34). *Regional utveckling – utsikter till 2020* [Regional development – Prospects until 2020].

SOU. (2008/09:93). *Mål för framtidens resor och transporter* [Objectives for future travel and transport].

Statistics Sweden. (2013). http://www.scb.se/sv_/Hitta-statistik/Artiklar/Antalet-barn-vantas-oka-i-Sverige/. Accessed 27 Sept 2017.

Statistics Sweden. (2016). https://www.scb.se/hitta-statistik/sverige-i-siffror/kommuner-i-siffror/#?region1=0380&region2=00. Accessed 29 Sept 2017.

Statistics Sweden. (2017). http://www.scb.se/hitta-statistik/statistik-efter-amne/miljo/markanvandning/tatorter-arealer-befolkning/pong/statistiknyhet/befolkning-i-tatort/. Accessed 27 Sept 2017.

Stenbacka, S. (2011). Othering the rural: About the construction of rural masculinities and the unspoken urban hegemonic ideal in Swedish media. *Journal of Rural Studies, 27*(3), 235–244.

Tillberg, K. (2001). *Barnfamiljers dagliga fritidsresor i bilsamhället – ett tidspussel med geografiska och könsmässiga variationer* [Families' daily leisure trips in the car society – A temporal jigsaw with geographical and gendered variations]. Doctoral dissertation, Acta Universitatis Upsaliensis.

Tillberg, K. (2002). Children's (in)dependent mobility and parents' chauffeuring in the town and the countryside. *Tijdschrift voor Economische en Sociale Geografie, 93*(4), 443–453.

Trafikanalys. (2015). *Uppföljningen av de transportpolitiska målen* [Follow-up on the objectives for transport politics]. Rapport 2015:7.

Trafikverket [The Swedish Transport Administration]. (2012). *Barns skolvägar* [Children's routes to school]. Report 2013:006.

Trivector. (2007). *Barns och ungdomars resvanor – en resvaneundersökning bland 6–15-åringar i olika stora orter* [Children's and young people's travel habits – A travel habit study among 6–15 year olds in different sized cities]. Rapport 2007:3.

UN. (1989). *The UN convention on the rights of the child.* http://www.ohchr.org/Documents/ProfessionalInterest/crc.pdf. Accessed 27 Sept 2017.

UN. (2015). *Transforming our world: The 2030 agenda for sustainable development.* https://sustainabledevelopment.un.org/post2015/transformingourworld/publication. Accessed 27 Sept 2017.

Uppsala Municipality Statistics. (2016). https://www.uppsala.se/contentassets/f09f9e6b994f41408c66064a2da8470b/omradesfakta-2016.pdf. Accessed 27 Sept 2017.

Uppsalatidningen. (2016, December 15). Framsteg ska sättas främst [Progress should be put first].

Urry, J. (2004a). The 'system' of automobility. *Theory, Culture and Society, 21*(4/5), 25–39.

Urry, J. (2004b). Connections. *Environment and Planning D: Society and Space, 22*, 27–37.

Valentine, G. (2004). *Public space and the culture of childhood.* Aldershot: Ashgate.

Westerstrand, J. (2008). Mellan mäns händer. *Kvinnors rättsubjektivitet, internationell rätt och diskurser om prostitution och trafficking* [Between the hands of men: Women's legal subjectivity, international law and discourses on prostitution and trafficking]. Uppsala: Uppsala University.

Whitelegg, J. (1993). *Transport for a sustainable future: The case for Europe.* London: Belhaven Press.

Zeiher, H. (2003). Shaping daily life in urban environments. In P. Christensen & M. O'Brien (Eds.), *Children in the city: Home, neighbourhood, community* (pp. 66–81). New York: Routledge.

Zeiher, H. J., & Zeiher, H. (1994). *Orte und Zeiten der Kinder. Soziales Leben im Alltag von Großstadtkindern [Places and times for children: Everyday social life of city kids].* Weinheim und München: Juventa Verlag.

Zinnecker, J. (1990). Vom Straßenkind zum verhäuslichten Kind. Kindheitsgeschichte im Prozeß der Zivilisation [From streetchild to the domesticated child: History of childhood in the process of civilization]. In I. Behknken (Ed.), *Stadtgesellschaft und Kindheit im Prozeß der Zivilisation. Konfigurationen städisher Lebensweise zu Beginn des 20. Jahrhunderts* (pp. 142–162). Opladen: Leske & Budrich.

Policy Material

The county plan for regional transport infrastructure 2014–2025 (Regional Federation of Uppsala County, 2014, part 1 and 2).

The regional bicycle plan for Uppsala county (Regional Federation of Uppsala County, 2010).

The action plan for the work with bicycle traffic (Uppsala Municipality, 2014a).

The action plan for public transport in Uppsala city 2015–2030 (Uppsala Municipality, 2014b).

CHAPTER 11

Gendering Mobilities and (In)equalities in Post-socialist China

Hilda Rømer Christensen

The image of China as a special case in terms of both transport and gender was formed during the Mao era, where the 'Kingdom of the Bicycles' became a signature for Maoist modernity and gender equality.[1] It was a period when the Chinese government took the lead in introducing the bike as an icon of communist universalism, equality and modernity, and launched a shared vision of a shiny bike, 'a Flying Pigeon in every Household' in parallel with the slogan that women held up half the sky (Rhoads 2012; Flying Pigeon n.d.). The ideal of the Kingdom of Bicycles of the Mao era was eventually connected to the nation-building process of the People's Republic of China (PRC) from 1949 onwards, a new beginning which included a radical strategy of equality between men and women in both legal and social terms. Female comrades (*nü tongzhi*) were encouraged and obliged to contribute to the rise of the nation and to work in the factories and fields. Both commercial and political propaganda in the twentieth century, interestingly, depicted women as sporty cyclists in picturesque landscapes and, later, as tractor drivers and Iron Girls.[2] The age of radical equality was followed by the Chinese reform era, starting in the 1980s, characterized by several, partially

H. R. Christensen (✉)
University of Copenhagen, Copenhagen, Denmark
e-mail: hrc@soc.ku.dk

© The Author(s) 2019
C. L. Scholten, T. Joelsson (eds.), *Integrating Gender into Transport Planning*, https://doi.org/10.1007/978-3-030-05042-9_11

mutually exclusive trends in gender discourse, with an essentialist turn emphasizing biology and natural gender differences that resurfaced during the 1980s and 1990s. Along with the developing market and consumer society, growing prosperity and the modern, well-educated career women (*zhí yè nü xìng*) of the twenty-first century, Chinese gender discourses have multiplied into a variety of ideals and practices. All in all, the varied displays of gender, femininity and masculinity in present-day China illustrate a breaking away from what is now perceived to be the conformity and gender neutrality of the Mao era (Yang 1999; Sudo 2007; Ko and Zheng 2007). Such trends have also surfaced in the new transport and mobility cultures, with varied displays and practices of new masculinities and femininities.

Reform policies and planning in China during recent decades have clearly been influenced by dominant Western paradigms of transport. These priorities have changed commuting modalities in Chinese metropolitan areas. This includes a steep rise in the use of private cars for commuting, along with a significant fall in the level of biking in urban China. According to estimates of the transport modal split issued by the Beijing Municipal Transport Commission in 2015, the use of cars has increased, their share of daily commuting rising from 5% in 1986 to 23.2% in 2000 and 34.2% in 2010. Biking, which used to be the dominant form of commuting, fell from 62.7% in 1986 to 38.5% in 2000 to 12.0% in 2015. In 2015, public transport, which has been revolutionized in terms of the provision of extensive metro and bus lines, reached a 50% share of daily transport modes, equally divided between metro and bus transport.[3] With respect to transport and car culture, critical environmental concerns have been mounting since the turn of the twenty-first century, addressing serious air pollution, and this has made issues of the low-carbon society and sustainable forms of transport vital. Making transport greener and smarter has become one cornerstone of Chinese politics and has taken effect in a number of efforts to curb the soaring number of cars, which account for 50–80% of CO_2 emissions in Chinese cities. Throughout, the public transport system has also been extended, and is intended to provide a new and efficient means of daily transport. In Beijing, for example, a total of 16 metro lines, together with a dense bus fleet were catering to millions of daily passengers in 2016 (Beijing Subway n.d.). This has affected daily mobility in urban China in both the use of public transport and the slow return of the bicycle as a daily mode of transport.

In this chapter, I will examine how social and gendered ideas and practices are embedded in these new forms of transport planning and mobility cultures and with what kinds of effects. The analysis will take a cross-cutting analytical look at gender and classed representations in car culture and public transport in Chinese metropolises. I will show that transport planning and mobility are intertwined in multiple ways with the emerging Chinese middle class and repertoires of masculinity and femininity, along with emerging gendered and social inequalities. Since transport and mobility have not been a central issue in gender studies and politics so far, the chapter also fills out a gap in knowledge production in the Chinese context and may provide an avenue for further research and policy interventions and provide a more inclusive transport and mobility culture.

WHAT'S THE PROBLEM: THE NEW MOBILITY PARADIGM

Over the past few decades, critical transport and mobility research in the West has applied many of the questions asked by sociologist Carol Bacchi in her influential work on the construction of policy problems (Bacchi 1999). She argues that dominant policy-making concepts are often marked by common sense; they apply a simple 'problem-solving' paradigm which assumes that 'problems' can be identified and that they are objective in nature. This paradigm, she contends, has become influential over the last few decades in the turn to evidence-based policy, including in so-called 'soft' areas such as education and health. In applying Bacchi's lens to the dominant field of transport, it seems obvious that transport research and planning have historically been characterized by a preference for hard evidence ever since the field's inception. Modern transport research took off during the period immediately following World War II and became influenced by the then optimistic belief in quantitative and rational (choice) approaches to rebuilding and developing post-war societies. In present-day Europe, transport policy—understood as planning, production and politics—still follows such hierarchical paradigms and principles of the post-war era; for example, in the top-down operation of city, regional and national transport planning, as well as in implementation processes that ignore the affordances and diversity of end users. This is an incremental model that tends to reproduce stereotypical ideas of technology, infrastructures and transport vehicles as a male-dominated domain. A side effect of this is that it operates with a uniform family and male-breadwinner model. The gap between planners, politicians and citizens has recently been recognized as damaging to the economy, growth and sustainability as well as to the wellbeing of citizens[4] (Levy 2013; Horelli et al. 2013; Horizon 2020).

In the following, I align myself with influential scholars in the field, such as Levy and Sheller, who have argued for the need to address and deconstruct core ideas of genderless models of choice and individuality as a way of introducing innovations into transport research (Sheller 2004; Hanson 2010; Levy 2013). One of the relevant discussions regards how it is possible to broaden transport research and to include more social identities as well as distributional and justice aspects into transport and planning procedures. For example, Caren Levy has argued along such lines for entanglements of transport and social identities and for what she calls the 'deep integration' of social identities in all areas of transport. Another related discussion has addressed the narrow 'rational choice' paradigm as inadequate for sustainable development or as a tool for creating inclusive transport and mobility. Sheller and Urry, for instance, focus on the links between transport and sustainability, and strongly challenge the prevailing individualistic 'choice and behaviour change' paradigm (Sheller and Urry 2016). They specifically question the *soft change* paradigm, which assumes that the practice of nudging rational actors can make individuals think and behave in a different (low-carbon) manner. The model of behaviour change that is based on the simple paradigm of ABC—an attitude, behaviour, choice approach—is too simple and not likely to create change in habits (Shove 2010; Schwanen et al. 2011). One example of the inefficiency of such an approach to sustainability might be the car licence-plate system practised in Beijing from 2008 meaning that on one day, all cars with a licence plate ending in an odd number were banned from driving, and on the next day it would be licence plates ending in an even number. The odd-even licence-plate driving restrictions were introduced during the 2008 Olympics in Beijing, restricting car use to days when the licence plate numbered odd or even was designated. Since then, the system has been changed several times (Schwankert 2014; Road Space Rationing in Beijing n.d.).[5] Rather than reducing the use of private cars, the measure simply prompted middle-class families to boost car consumption by buying a second car with a differently numbered licence plate, so that they would have at least one car allowed on the roads every day.

Sheller has instead suggested a different approach, an 'emotional turn' in the study of how people get around, to what has been called 'quotidian mobility'. Paying qualitative attention to emotions and feelings that are embedded in patterns of daily mobility, she argues, could help to shift attention away from the counter-factual 'rational actor', who is supposed to make carefully reasoned economic choices. Following Sheller, research should rather look at the lived experiences of dwelling with cars and other modes of transport in all their complexity, ambiguity and contradiction; and making such themes vital fields of study (Sheller 2004: 222). In this chapter, I will demonstrate how living not only with cars but also with other modes of transport has become a vital part of everyday (middle-class) life in urban China. It turns out that different forms of mobility also represent various emotional and embodied landscapes and experiences, which are feeding into transport planning and policies. I will show that gender and class are implied and are co-constructing such landscapes, and are also implied in the potential avenues towards change. In addition, Miller argues that cars and mobility must be addressed in a broad sense as a culture embedded in intimate relationships between cars and other forms of transport and people (Miller 2001: 17) So far, such aspects have been marginal in transport research, both in China and in the West. Both emotions and the reflection of various social categories and end users have mainly been left to commercial agents and branding strategies to address. Such analyses, as I will show in the section on car culture, have obviously more often than not been serving commercial aims and have acted as unstable allies for sustainability, justice and the common good.

Key Concepts

In order to take forward these ideas of transport, class and gender as they are implied in the current making of Chinese urban life and culture, I will simplify the theoretical considerations into a couple of cross-cutting concepts. First, the idea of global *assemblage(s)* seems to provide a useful analytical lens for exploring the formation of current post-socialist modernity and the new global cultures of mobility (Salter 2013). The idea of a global assemblage is suitable for examining how social relations have been intertwined with material objects at multiple levels, not least when it comes to the creation of new modalities of transport and the powerful position of car culture in China and around the world. A global assemblage, according to anthropologists Zhang and Ong, is marked by a particular 'global' quality and refers to phenomena that are attractive, mobile and dynamic, that move across analytical borders and at the same time reconstitute 'society', 'culture' and 'economy' in their known or imagined forms (Zhang and

Ong 2008). The notion of an assemblage moves beyond deterministic concepts in both neo-Marxist and post-structuralist approaches; for example, the often-exaggerated use of hegemony in the Gramscian sense, and the power/knowledge complex in Foucault's idea of the dispositive. Salter, for example, has identified the circulation 'assemblage', which galvanizes understandings of the dynamic co-constitution of mobile subjects and the deep structure of mobility (Salter 2013). Cars and new forms of mobility clearly illustrate both such connected processes and their material forms, which have changed the horizon of society and daily lives (Christensen 2015; Sheng 2015). In the context of present-day China, we need only think of the exploding car market, elevated highways, clean and extensive metros, high-speed trains and new airports that connect urban centres. Another, and softer, field is related to the field of new biking. Cycling technology too is in continuous flux in the wake of both mass and specialized production. Bike production, like car production, has become part of a new integrated global system of production, a cross-sectional, post-Fordist system linked to media and to new modes of consumption and governance.[6] That said, could one pursue these lines of thought to address cycling culture as forming a different and possibly more sustainable and more inclusive form of global assemblage? I will show how both types of assemblage enable the emergence of new gendered subjectivities and new social forms, while also creating new social constraints and inequalities.

These considerations are linked to my second key notion, the idea of *interpellation*, which sharpens understandings of how technologies and subjectivities are co-constituted in the use and appeal of material artefacts. Interpellation as a concept derives from the French philosopher Louis Althusser and refers to the idea of 'hailing' as a kind of subjectivation of a person into her or his social and ideological position by an authority figure.

The concept of interpellation has also been taken up in broader arenas of cultural studies in recent studies of cars and masculinity (Lees-Maffei 2002; Balkmar 2012; Landström 2006). Interpellation, according to the Swedish sociologist Catharina Landström, for example, allows cars to be seen as artefacts that construct and enable gendered subjects. Certain processes of interpellation, she argues, initially invite men into an imagined homo-social community and a shared culture of transport artefacts, which sustain gender differences and masculine norms (Dant 2004).[7] Car culture is therefore implied in a certain way of doing heterosexual masculinity and pleasure. In such a stereotyped framework, women become constructed as practising a rational femininity as opposed, or even as a threat, to this type of male sociality and pleasure.[8] The following analysis will show how varied types of transport and mobility practices interpellate women and men in different ways in present-day Chinese transport culture. It is argued that the reflections of such processes of interpellation are vital if we are to gain deeper understandings of current Chinese transport strategies.

DATA MATERIAL

I have used articles from Chinese newspapers as vital data for the analysis in this chapter: *China Daily* and *Shanghai Daily* in English, and *Renmin Daily's* Chinese version. Chinese newspapers are routinely regarded as merely tools of government policy and not as forums for reflections or critical debate (Shirk 2011). Yet, in the case of transport and mobility, these newspapers do provide a relevant channel for locating directions in government policy and new understandings of mobility and the Chinese middle class.[9] They not only offer governmental interpretations of forms of mobility, in particular the introduction of car culture associated with the rising middle class. The English versions also provide laboratories or so-called trial balloons for floating new directions and interventions in Chinese transport policy. Nevertheless, the attention given to motorized and non-motorized mobility, private and public transport has been very uneven in all three newspapers. This also points to the often-paradoxical priorities of Chinese transport policies—the goals of making it greener and smarter and introducing low-carbon policies on the one hand, and the hegemony of growth and partnership with the car industry, and powerful government interests in these, on the other.[10] At the present moment, the Chinese press seems to speak in two complementary discourses. On the one hand, the English-language newspapers reflect quintessentially middle-class values and celebrate the individual and consumer ambitions of the urban middle class. The Chinese edition of *Renmin Daily*, on the other hand, seems to maintain and revitalize the emphasis on collectivity and the nation, but with an unambiguous stress on motorized transport seen in the early sentiments of a 'virtuous circle of infrastructure and car consumption'. *Renmin Daily* also repeatedly stresses how car consumption 'can change society, promote social civilization and become an effective tool for progress' (*Renmin Daily*, 5 January 2004; *Renmin Daily*, 27 August 2009). This is a horizon which has by and large remained intact in the media as well as among metropolitan residents in their daily practices, in spite of serious problems of transport inefficiency and air pollution.

The data material was collected through thematic screenings of online newspaper archives dated from 1995 until 2017. The *China Daily* archives provide the most comprehensive material, while the online archives of *Shanghai Daily* from 2005ff. and the Chinese-language *Renmin Daily* from 2009ff. are more selective.[11] For analytical purposes, I have identified a cluster of around 20–40 core articles in each newspaper that address relevant representations of gender in terms of policy, materiality and individual practices. In order to validate the media representations, I have tested findings and analytical perspectives through fieldwork in car showrooms and bike shops in Beijing and Shanghai.[12] I also conducted a mini survey in Shanghai in 2013 with around 240 respondents, which included questions on daily practices and gendered relevant values across the three main types of transport: cars, bikes and public transit.[13]

The empirical data was connected and analysed by combining Norman Fairclough's methodology of critical discourse analysis with the operational approach of an emotional sociology of mobility (Fairclough 1995). Following Sheller's idea of emotional sociology in transport research, I include three aspects: (1) the preferences of individual drivers and passengers, (2) the meso level of located and changing transport cultures and (3) the macro-level pattern of regional, national and transnational emotional/ cultural and material geographies. Fairclough's idea of critical discourse analysis implies a three-tiered methodological strategy focusing on text, discursive practice and social practice. Text in this context refers to the textual analysis of newspaper articles, while discursive practice refers to the ways in which transport is talked about, and social practice indicates the broader social formation. The themes that are the focus of the following paragraphs are the 'material bricolages' that emerged out of exploratory thematic and cross-cutting screenings and combinations of repeated screenings of the newspaper articles and the collected data. All aspects have been merged into an analysis of intertextual structures and overall ways of representing mobilities, gender and class in post-socialist China.

WOMEN CAR LOVERS

The white-collar segment that represents China's rising middle class of men and women has been of vital importance in the ongoing marketization and promotion of car culture in China. It is interesting that both media and market here have been active in exposing the image of a new gender-equal family ideal. Hence, a central target in car branding has been the well-educated couple, consisting of an outgoing career woman and a soft or metrosexual man (Hird 2011; Xiao 2010a). At the same time, the media exposure reveals highly gendered patterns of representation. In *China Daily*, female car drivers are constructed as a new type of woman, who is described as feeling a close affinity and love for her car. This discourse tends to present women as having emotions towards cars and taking pleasure in them.

> It has been a long day at work, but the 27-year-old lawyer Li Liali is about to meet a dear friend in the parking lot of North Beijing Guohua Plaza, which is her red Buick regal. The woman's exhausted expression melts into a grin as she slides into the driver's seat. She fires up the engine, wraps the seatbelt around her, turns on the GPS tracker and shifts into gear. (*China Daily*, 16 November 2010)

The female car lover, as noted by Balkmar (2012: 138), may be regarded as an emerging symbol of women and cars, and as a figure who challenges the dominant ideas of men's monopoly of car pleasures. Such gendered interpellations moderate popular wisdom, which attributes to men and masculinity the evils of car culture and the idea of cars as a kind of metallic phallus linked to men and bodily desires (Landström 2006). Female drivers are indeed portrayed as drivers, yet they are often framed into a heteronormative context, where their interests and pleasure become romanticized and women's car driving becomes identical with care work—and not with their own individual aims and pleasures. What is more, the bold efforts in Chinese car branding have primarily been directed at married middle-class couples and have made car producers more sensitive to women's preferences. The car industry currently applies a variety of class and gendered branding strategies in recognition of shifts in values and growing equality when it comes to car purchasing among families. This is visible in a shift in colours, from the elite black cars to more colourful and softer forms designed to match the preferences of Chinese consumers, notably white-collar families in big cities. The growing popularity of so-called compact cars, in contrast to pretentious sedans, may tap into a more gender-neutral taste (Christensen 2015).

Here, the Mini Cooper features as an exception that addresses single women. This car, which was launched in Britain as part of Western youth culture in the 1960s, has become the leading car in the mini luxury segment in China. In 2013, 80% of Mini cars in China were owned by women.

Hence the Mini was presented as being imbued with a specific appeal for women, and this is transposed into commercial slogans and narratives that address single women of a mature age (Xiao 2010b). 'I am already over 60, but I would like to have a car that shows the real feelings in my heart', says the fictional Auntie Chen, who is named as one of the first owners of a Mini car in China. Mini car-owners in China are described by the Mini Cooper marketing director as, 'fashionable women who enjoy good-looking things and usually don't care whether a car can seat four adults or just her and her pet' (*China Daily*, 23 December 2010).

The Mini at this time seemed to be challenging the Chinese ideal of a family car by celebrating the individuality of the well-off single woman. In this way, the Mini is also in the business of co-producing the image of an independent lifestyle among a certain group of women in present-day China and making the group of well-off seniors appear younger and more attractive for branding strategies. Their practices and self-perceptions present a far cry from the victimizing ideas of well-educated single women blamed by the Chinese public for being 'leftover women' (*Shèng Nǚ*) due to their unmarried status (Fincher 2014).

EDUCATED MEN AND SUPER-DRIVERS

Studies of Chinese masculinity have depicted the multiple forms of masculinity that exist, and have revealed how notions of masculinity past and present circulate in contemporary culture and debates. In a pioneering study, Kam Louie has claimed that the *Wen-Wu* 文武双全 dyad in Chinese tradition is a paradigm that explains the performance of gender identities in particular masculinity (Louie 2002: 4; Hird 2011).[14] The *Wen-Wu* dyad suggests an intra-acting and inclusive soft-hard masculinity, with *Wen* associated with cultural attainments and *Wu* with martial abilities. In transposing this framework of the *Wen-Wu* dyad into the twenty-first century, two distinct representations of urban middle-class men and of masculinity and cycling emerge. The *Wen* side is represented by the New Chinese white-collar male, a feminized, emasculated man who has been placed at the symbolic heart of China's economic success (Barries 2013). The other side of the dyad has resurfaced with the *Wu* characteristics embodied in the new entrepreneurial masculinity, connected with aggressiveness and strength.

The most expressive trend in Chinese car culture, as presented by the media, not only links to the martial, *Wu* side of the repertoire, it also connects to a global idea of masculinity in which the imagined needs and pleasures of the male business and government elite have been actively incorporated. This can be seen in the launch of longer and more pretentious luxury car models that are assumed to attract Chinese (male) consumers who like to move around in chauffeured cars (Dunne 2011; Christensen 2015). What is exposed is the conspicuous and adventurous car consumer, who seems to accentuate an updated manifestation of *Wu* masculinity; for example, in marketing reports presented for *China Daily* readers. There is, for example, a McKinsey report from 2012 on luxury car consumers, which nurtured notions of aggressive consumerism and strong masculinity. The report bluntly described Chinese customers as 'obsessed with presenting a successful image', and many younger male buyers were portrayed as seeing the car as a 'business card' signalling their credibility. The report also presented differences, carved out along gendered lines, with male buyers favouring 'socially recognized' premium brands, in contrast to female buyers, who 'put a priority on exterior styling, safety features and comfort'. Another report along the same lines divided China's luxury car buyers into five segments. A majority of these customers were identified as younger males, while women made up a distinct and more mature group among business and executive purchasers.[15] The other side of the repertoire, represented in the softer *Wen* values, have recently been studied by Hird (2011), who focus on the new white-collar masculinity, which has been said to be central and at the symbolic heart of China's economic success (Hird 2011). This mainstream middle-class masculinity, assumed to be the new normal and most common form, tends to disappear and be over-layered by new tough masculinity in the branding strategies and media representations.

The available sociological evidence of gendered preferences and practices is scattered and uneven when it comes to gender and class assumptions about car preferences. My own survey from late 2013, which involved 240 respondents in the Shanghai metropolitan area, provided unstable sociological evidence of consumer preferences and gender similar to Landström's (2006) findings. Most of the respondents, irrespective of gender, identified the family car as their dream-car and not the luxury sedans that marketing gurus, the business community and car producers assumed they would want. The desire for a family car reflects the central position held by the Chinese family, and resembles the social dimensions of car culture, the car as a family member, as claimed by Sheller (2004).

Sociological evidence points to a surge in the numbers of both male and female drivers, yet the gendered inequalities in driving practices seem to persist in China. In 2013, the number of private cars in the country reached 85 million, compared to only 6 million ten years earlier. While men took the lead as drivers at the turn of the millennium, in recent years the increase in the number of women drivers has accelerated, and they are now registered both as drivers and owners. Between 2003 and 2013, the total number of women with a driver's licence increased massively from 20 million to 60 million. Officially, women today make up 40% of Chinese car-owners, compared to only 25% in 2003 (Xiao 2019).[16] However, these surging numbers may conceal attempts to bypass new government regulations restricting car ownership and driving in some city areas. Nevertheless, women car drivers still make up a minority, at 22%, of all Chinese drivers, a fact which tends to leave women with more sustainable, yet also more limited, travel choices compared to men.

PUBLIC TRANSPORT: GENDERING THE COMMUTER BURDEN

Whereas road construction and individual cars were prioritized during the first decades of the post-socialist period in the 1980s and 1990s, more emphasis and resources have been allocated to public transport, in particular railways, in recent years. For example, in 2012, investment in public transport was five times that in road construction in Beijing (Zhao and Li 2016: 947). More generally, from a political angle public transport has been regarded as a common good and a provision which directly affects the everyday lives of residents of a city. The provision of public transport—or the lack thereof—can have a severe impact on class formation and gender relations, and this is clear in the case of public transport provision in present-day urban China. While the public transit system is in principle democratic and open to all, it seems as though existing transport modalities in Beijing primarily seek to serve the needs and pockets of the middle and higher stratum of the middle class (Zhao and Li 2016: 955). Not only pricing policies, but also planning and investment in the Beijing Subway exhibit pronounced middle-class values and aesthetics, which also have gendered side effects. On the surface, the new and still sprawling subway system, carriages, stations and hallways ideally represent an ordered and clean world devoid of embodied marks of smells, bodily tissue and noise. Spitting, urinating, belching and coughing in public areas, such as the metro stations, are regarded as something belonging to the past blue-collar world of the Mao era. Passengers are mainly young, urban, white-collar workers and students, with young and fit bodies, who are able to navigate the maze of escalators and often long and strenuous walks between the different lines and destinations.

Ideally, the metro represents a transport mode with evident advantages, such as 'speed, capacity and immunity from bad weather' (*China Daily*, 16 June 2009). Amidst such shiny images of seamless and convenient metropolitan transport, the middle class in the big Chinese cities has its own commuter burden, which also seems to hit women disproportionately due to their lower level of car ownership and driver's licences and lower salaries. Indeed, metro commuting has become a new daily exercise in survival, as witnessed by a young Shanghai woman.

> My typical working day starts with a struggle in the metro train. Since I work in the CBD area and can't afford to live close by, I need to change between two metro-lines, the two busiest in Shanghai. The metro system is full of unpleasant surprises every day. Delays and inadequate air conditioning are often not the worst problems; the worst is when you try to get off the train, and hundreds of impatient passengers from outside can't wait to rush in. Dirty shoes are just one of the bitter results of such near stampedes. (*China Daily*, 5 August 2010)

The constraints and inconveniences of daily transport undoubtedly contribute to the widespread lack of happiness among residents of both Beijing and Shanghai, as reported in 2010.[17] During rush hours, metro carriages are stuffed with standing and mute individuals, who struggle to use their mobile phones as a refuge during the long, strenuous rides. This is a fact which is also reflected in the media reports and the growing recognition of 'security problems' on the new public transport system. 'The Subway is so crowded at peak times many passengers need to be pushed by subway station employees to get in the carriage' (*China Daily*, 16 May 2009). Over the last decade, subway transit has plunged down the value ladder and become one of the least popular transport modes, due to overcrowding, underground air pollution and safety issues, and marked by 'unbearable stress' and 'latent danger'.[18] In spite of the bold extensions to the subway network in Beijing, it seems that there are limits to a system which cannot absorb the millions of daily commuters in this mega city. The event of raising prices on the Beijing Transit system in 2014 constitutes a revealing case in point. The transport authorities in this case argued that the existing cheap and flat-rate system should be changed towards raising prices as a step towards a more equal economic redistribution. Since most users were white-collar employees going from the suburbs to the city centre, they ought to pay more. 'A low-price strategy is not a long-term solution' as one of the experts at the Ministry of Transport said, 'we need to rate the public transport system through comprehensive indexes, such as *comfort, convenience and safety*' (*China Daily*, 21 July 2011, MU.) At the same time, the Beijing Commission of Development and Reform initiated a bottom-up process and launched a survey which over a few weeks spurred more than 40,000 suggestions and comments from about 24,000 citizens. Many respondents were addressing the issue in a broader context than the authorities, and envisaged a variety of ideas of class and justice and equal distribution of resources.[19] Yet, the users on social media did address the unequal social and gendered consequences of abandoning the former cheap and flat-rate system. Less well-off women lamented being left with limited and unaffordable choices. For example, one young mother calculated her annual transportation cost to be 2112 RMB and compared it to the price of three cans of imported milk powder which is regarded as a Western luxury item due to high import taxes. A female cleaner reported a rise in her annual transportation cost to 2640 RMB, which equalled the annual tuition fee of her daughter's school. A third woman reported her rising transportation costs to be over 3000 RMB, which came to half of her monthly salary of around 6000 RMB. Several commentators also saw the subway as a social provision for the poor: 'Please do not raise ticket prices! As the most efficient and fair traffic tool, the subway [needs] to provide more access for low-income people'. And one resident even suggested a progressive tax rise: 'Rich people should pay higher taxes to subsidize traffic transport. The current ticket price in Beijing is important for low-income people. Please don't let us perceive Beijing as a cold city'. A survey from 2014 echoed these individual consequences in showing the substantial inequality inflicted by Beijing planning and transit. The daily commuter burden turned out to be heavier for the low-income group—and for women in this group in particular (Zhao and Li 2016: 955).

In general, both male passengers and male car drivers were represented, but in a different dialogue related to the rising costs of parking and taxis rather than to the ticket price of public transport. An example is Li Ming, who could not afford a daily parking fee of 70 RMB Yuan and therefore, like many drivers, had to switch to less efficient public transport five days a week. 'But Li, who used to drive for about 45 minutes from home to work, now spends about 80 minutes on the subway and then a bus' (*China Daily*, 12 April 2011). Others were reported to have found alternative solutions, such as carpooling or private shuttle buses. Di Zhenxing took a crowded bus to work for several days, which he described as 'a horrible experience'—and so he organized a car pool with his neighbours to reduces the cost of parking but maintain the convenience of travelling by car (*China Daily*, 12 April 2011). A third man changed from his car to a private shuttle bus to avoid the metro: 'The subscription shuttle bus offers me a point-to-point service. It is clean and comfortable and better than the crowded metro' (*China Daily*, 18 October 2013).

The Beijing survey opened up a plethora of 'suggestions' that revealed transport as being entangled in much more than just moving from A to B. As pointed out by Levy (2013), transport 'choice' pertains to wider issues of identities, governance and justice. The survey concluded with a price rise, which accommodated most of the presumably better-off responders, while the approximately 30% who lamented the price rise were ignored. So, deliberations and so-called bottom-up approaches via social media are no guarantee of social or gender-balanced solutions.

The pace of Chinese transit developments taps into the ongoing global competition, with China aiming to become the world's number one in this field. The *Renmin Daily* trumpets the success of the Chinese transit developments compared to the West (with the refrain of pace, whereby it took London's Underground system 150 years to develop what Beijing and Shanghai have accomplished in 20 years in terms of metro lines). Nevertheless, it seems that the Beijing transport plans for 2030 have not come up with long-term or innovative solutions, but rather operate with more of the same: new metro lines in a sprawling and ever-larger Beijing metropolitan area may serve the agenda of global competition and national pride, which ignores the daily costs for a growing number of Beijing residents, not least women, who will be burdened with more and longer commuting.

CONCLUDING REFLECTIONS: CONCEPTUAL AND STRATEGIC

China right now seems to represent both mounting challenges and potential for change in the field of transport and gender. Beginning with the idea of global assemblages, I have demonstrated how new forms of mobility have broadened the horizons of society and daily lives in urban China. The exploding car market, elevated highways, clean and extensive metro systems and changing urban infrastructure have enabled the emergence of new social and gendered subjectivities and new institutional and social forms, while also creating new constraints and inequalities.

More specifically, I have demonstrated how new hierarchies—gendered and (dis)embodied—have been produced through media representations and social practice. Both implicit and explicit transport planning forms the backdrop for the prevalent assumptions and priorities. In returning to Sheller's approach of emotional sociology and living with the various transport modes, there seem to be clear differences in the representation of various forms of transport at the current stage of post-socialist development. On the one hand is the enthusiasm and pleasure transmitted via media reports of car culture and leisure cycling, which are often presented in both material and embodied ways: as a shiny new car or a smart bike and gendered male or female bodies (Christensen 2015; Christensen 2017). At an overall level, this contrasts with the representation of public transport, which seems disconnected from bodies and pleasure. The new metros and extended bus lines, for instance, lack the investment of positive embodied imaginations and desires, which turns out to be a vital and as yet invisible element when it comes to 'choice' and 'preferences' in transport planning and practices. A precondition for this is to combine both pleasure and functionality that appeal to everyone, across gender, age and class, as well as to different types of transport.

At the strategic level, it has been widely assumed that women engage in more sustainable mobility practices than men, and several studies have provided evidence for such a belief, in findings which recur in different parts of the world (Transgen 2007).[20] For instance, there seem to be enduring gender differences, often to the advantage of women, as well as sustainable practices, in their use of non-motorized rather than motorized transport, not only in China, but also according to cross-national studies (Zauke and Spitzner 1997; Polk 2003; Rosenblom 2006; Johnson-Latham 2007; Srinivasan 2008).

Yet these particular practices and assumptions also constitute a political and conceptual challenge for transport planning and political priorities. If, for example, women's more sustainable mobility is a reaction to the restrictions they experience and inequality of access, then we are confronted with a delicate paradox of equality and sustainability: enhancing gender equality within the present transport paradigm, in which cars are prioritized, would imply more women drivers, which would increase women's carbon footprint. Another strategy would be to urge men to change their mobility practices and to use women's mobility as a model, a suggestion which it would be very difficult to implement. The paradox involved invites both the posing of new research questions and the development of a more robust knowledge base in various fields of transport planning and policies.

This paradox might also be resolved through the introduction of new transformative modes and practices of transport. Here, new technologies such as bike-sharing might become part of the solution if they were combined with public interventions and strategies of social justice. The current situation in China, as in many other parts of the world, is that existing transport systems, with cars given priority, the inadequate provision of public transport and various new modes of private cycling, tend to leave women with more limited opportunities and choices than men. At the end of the day, this might prove to be detrimental to society, market potential and the quality of life for all.

NOTES

1. The bike was one of the so-called Big Four commodities, on a par with the sewing machine, the television and the washing machine, which signalled that a Chinese family was 'modern'.
2. The sporty and mobile women were depicted in propaganda posters and postcards. While there was just over one bike per urban household in 1978, the rate was more than two by 1984 (Rhoads 2012). Since most households tend to first provide bikes and transport tools for men, I assume that women were catching up as bike owners and cyclists at the beginning of the transition period in the 1980s.
3. Green Transportation model split, Beijing, 2015.
4. In the powerful European transport research there also seems to be a growing recognition that existing transport models are preventing more innovative and inclusive processes (Horizon 2020/AG coping paper 2014, Smart mobilities, Bruxelles 2016).
5. Since 2008, the system has been changed so that drivers may use their car every business day except one, as two licence-plate final digits are restricted per day. At weekends, no restrictions apply (Schwankert 2014; Road space rationing in Beijing n.d.).
6. Huybers-Withers, 'Mountain Biking is for Men', 1208.
7. A sensitive but gender-blind claim is pursued by Tim Dant (2004), who posits the assemblage 'driver-car' both as a (genderless) product of human design, manufacture and choice and as an enabler of social action that has become routine and habitual and which affects many aspects of life in late modern society.
8. Interpellation is a contested but often-used concept in gender studies, notably in the post-structuralist versions. Performativity is a related concept suggesting that processes of interpellation are enacted according to gendered scripts and gendered economies of pleasure expressed in the concept of a heteronormative gender matrix, cf. Butler (1999).

9. The data material has been provided by thematic screenings of online newspaper archives: *China Daily* 1995 ff., and *Shanghai Daily* 2005 ff., plus paper editions of *China Daily* 1985 ff. and *Renmin Daily*'s online archive in Chinese from 2009–2016.

10. In *China Daily*, up until 2013 the number of articles dealing with bikes came to just over 300, while car-related articles came to nearly 20,000. In *Shanghai Daily*, bike-related articles totalled a mere 128 as against thousands of articles on cars, often marketing specialized car brands. *Renmin Daily* in Chinese represents the same overall pattern, with a total of 565 articles on bikes, while articles on cars numbered nearly 9000.

11. Core articles consist of a variety of lengthy feature articles, reports on leisure and club life, news articles on the consequences of change in transport policy and occasionally the opinions of journalists and commentators. Many articles in the English-language press contain a broad range of facts supplemented by interviews with experts and residents, while *Renmin Daily* in the Chinese language is dominated by politicians and experts.

12. The fieldwork consisted of interviews with bike dealers and customers and was conducted in May–June 2014. I visited ten bike stores, five in Beijing and five in Shanghai. All the stores were located in what have become specific bike store areas, such as in and around Jiao Daskou East Street, Beijing, and bike shops close to Tsinghua University. The Shanghai bike shops were located in and around Jian Guo Road in the French Concession area.

13. This survey was conducted to fill the gap of gender-sensitive statistical data in present-day China. Detailed consecutive surveys such as *Chinese Women's Social Status* (1990–2000, 2010) do not include transport. Similarly, general analyses, such as household analyses, do not specify transport and daily mobility in the household or related to gender.

14. Most of Louie's points here are referred from Hird (2011).

15. 19 June 2013 in *China Daily*, Li Fangfan is here quoted from *People's Daily* online 10 June 2013.

16. *China Daily* 19 June 2013.

17. Shanghai and Beijing came out at the bottom among 35 surveyed cities regarding middle-class happiness. Study: middle-class families in Beijing and Shanghai are in pseudo-happiness. In *Renmin Daily* online, 17 March 2010. The survey, which covered more than 70,000 residents between the ages of 20 and 40 in ten cities nationwide, implied that people aged between 30 and 35 located in smaller cities were the happiest. Note that daily commuting in heavy traffic was mentioned as one major reason for the discontent, along with competition at work, and the high cost of children's education.

18. For example, Line 4 in 2013 was reported to carry one million passengers per day, and the transfer stations of lines 1 and 2—the oldest in the network—were reported as being under 'unbearable stress' and as a 'latent danger' (*China Daily*, 16 September 2013 & 16 June 2009).

19. The rising prices of the Beijing Subway were commented upon online in a survey issued by The Beijing Commission of Development and Reform and *Renmin Daily*, which spurred 40,222 comments made by 24,079 persons. Many of these addressed far more than just the increase in ticket prices!

20. List from Hanson (2010).

REFERENCES

Bacchi, C. L. (1999). What's the problem represented to be? In C. L. Bacchi (Ed.), *Women, policy and politics: The construction of policy problems* (pp. 15–50). London: Sage.

Balkmar, D. (2012). *On men and cars: An ethnographic study of gendered, risky and dangerous relations*. Dissertation, Linköpings Universitet.

Barries, M. (2013, March 22). New Chinese drivers Crave Luxury and prestige. *China Daily*. http://usa.chinadaily.com.cn/epaper/2013-03/22/content_16335498.htm. Downloaded 2017-08-07.

Beijing Evening Paper (2015, January 23) & online survey.

Beijing Subway. (n.d.). *Wikipedia*. https://en.wikipedia.org/wiki/Beijing_Subway. Downloaded 2016-01-03.

Butler, J. (1999). *Gender trouble: Feminism and the subversion of identity*. New York: Routledge.

Christensen, H. R. (2015). The lure of car culture: Gender, class and nation in 21st century car culture in China. *Women, Gender and Research, 24*(1), 110–123.

Christensen, H. R. (2017). Is the kingdom of the bicycles rising again. Cycling, gender and class in post – Socialist China. *Transfers. Interdisciplinary Journal of Mobility Studies, 7*(2), 1–20.

Dant, T. (2004). The driver-car. *Theory, Culture and Society, 21*(4–5), 71–79.

Dunne, M. J. (2011). *American wheels Chinese roads: The story of general motors in China*. Singapore: Wiley.

Fairclough, N. (1995). *Critical discourse analysis: The critical study of language*. Boston: Addison Wesley.

Fincher, L. H. (2014). *Leftover women: The resurgence of gender inequality in China*. London: Zed Books.

Flying Pigeon. (n.d.). *Wikipedia*. https://en.wikipedia.org/wiki/Flying_Pigeon. Downloaded 2013-10-13.

Hanson, S. (2010). Gender and mobility: New approaches for informing sustainability. *Gender, Place & Culture, 17*(1), 5–23.

Hird, D. (2011). *Anxious men and the recuperation of masculinity in contemporary China*. Dissertation, University of Warwick.

Horelli, L., Jarenko, K., Kuoppa, J., Saad-Sulonen, J., & Wallin, S. (Eds.). (2013). *New approaches to urban planning: Insights from participatory communities*. Aalto: Aalto University.

Horizon. (2020). https://cordis.europa.eu/project/rcn/218633/factsheet/en and http://www.systematica.net/news/systematica-kicks-off-diamond-project/

Johnson-Latham, G. (2007). *A study on gender equality as a prerequisite for sustainable developments*. Report to the Environment Advisory Council. Sweden: Ministry of the Environment.

Ko, D., & Zheng, W. (2007). Translating feminisms in China. In D. Ko & W. Zheng (Eds.), *Translating feminisms in China* (1–13). Oxford: Blackwell.

Landström, C. (2006). A gendered economy of pleasure: Representations of cars and humans in motoring magazines. *Science Studies, 2*, 31–53.

Lees-Maffei, G. (2002). Men, motors, markets and women. In P. Wollen & J. Kerr (Eds.), *Autopia: Cars and culture* (pp. 363–270). London: Reaktion Books Ltd.

Levy, K. (2013). Travel choice reframed: 'Deep distribution' and gender in urban transport. *Environment and Urbanization, 25*(1), 47–63.

Louie, K. (2002). *Theorizing Chinese masculinity: Society and gender in China*. Cambridge: Cambridge University Press.

Miller, D. (2001). Driven societies. In D. Miller (Ed.), *Car cultures* (pp. 1–33). Oxford: Berg.

People's Daily (Renmin Daily) China Daily. (2013, June 19). Li Fangfan here quoted from *People's Daily* online, June 10.

Polk, M. (2003). Are women potentially more accommodating than men to a sustainable transport system? *Transportation Research: Part D, 8*, 75–95.

Rhoads, E. (2012). Cycles of Cathay: A history of the bicycle in China. *Transfers, 2*(2), 95–120.

Road space rationing in Beijing. (n.d.). *Wikipedia*. https://en.wikipedia.org/wiki/Road_space_rationing_in_Beijing. Downloaded 2017-08-07.

Rosenblom, S. (2006). Understanding women's and men's travel patterns: The research challenge. In *Research on women's issues in transportation*. Volume 1, Conference overview and plenary papers, conference proceedings, Vol. 35, pp. 7–28.

Salter, M. (2013). To make move and let stop: Mobility and the assemblage of circulation. *Mobilities, 8*(1), 7–19.

Schwanen, T., Banister, D., & Anable, J. (2011). Scientific research about climate change mitigation in transport: A critical review. *Transportation Research: Part A, Policy and Practice, 45*(10), 993–1006.

Schwankert, S. (2014, November 27). Beijing considers permanent odd-even license plate restrictions. *The Beijinger*. [Blog]. http://www.the beijinger.com/blog/2014/11/27/beijing-considers-permanent-odd-even-license-plate-restrictions/. Downloaded 2017-08-07.

Sheller, M. (2004). Automotive emotions: Feeling the car in automobilities. *Theory, Culture and Society, 21*(4–5), 221–242.

Sheller, M., & Urry, J. (2016). Mobilizing the new mobilities paradigm. *Applied Mobilities, 1*(1), 10–25.

Sheng, Q. (2015). Spatial 'complexity': Analysis of the evolution of Beijing's movement network and its effects on urban functions. *Footprints: Delft Architecture Theory Journal. Theme: Mapping Urban Complexity in an Asian Context,* (2), 32–42.

Shirk, S. (Ed.). (2011). *Changing media, changing China.* Oxford: Oxford University Press.

Shove, E. (2010). Beyond the ABC: Climate change policy and theories of social change. *Environment and Planning A, 42*(6), 1273–1285.

Smart Mobilities, Conference Proceedings. Bruxelles 2016.

Srinivasan, S. (2008). A spatial exploration of the accessibility of low-income women: Chengdu, China and Chennai, India. In T. P. Uteng & T. Cresswell (Eds.), *Gendered mobilities* (pp. 159–171). Burlington: Ashgate.

Sudo, M. (2007). Concepts of women's rights in modern China. In D. Ko & W. Zheng (Eds.), *Translating feminisms in China* (pp. 13–35). Oxford: Blackwell.

Transgen. (2007). *Gender mainstreaming European transport research and policies: Building the knowledge base and mapping good practices.* (EU report by H. Rømer Christensen, H. Hjorth Oldrup & H. Poulsen). Koordinationen for Kønsforskning, Københavns Universitet.

Xiao, C. (2010a, September 16). Steering the market. *China Daily.* http://www.chinadaily.com.cn/life/2010-09/16/content_11310323.htm. Downloaded 2017-08-07.

Xiao, X. (2010b, December 23). Mini branding for youthful market and young at heart. *China Daily.* http://usa.chinadaily.com.cn/epaper/2010-12/23/content_11745492.htm. Downloaded 2017-08-07.

Xiao, M. (2019). Annual report on the trend of automobile consumption in China (2012). In *Development of a society on wheels* (p. 285). Singapore: Springer.

Yang, M. M.-H. (1999). From gender erasure to gender difference: State feminism, consumer sexuality, and women's public sphere in China. In M. M.-H. Yang (Ed.), *Spaces of their own: Women's public spheres in transnational China* (pp. 35–67). Minneapolis/London: University of Minnesota Press.

Zauke, G., & Spitzner, M. (1997). Freedom of movement for women: Feminist approaches to traffic reduction and a more ecological transport science. *World Transport Policy and Practice, 3*(2), 17–23.

Zhang, L., & Ong, A. (Eds.). (2008). *Privatizing China, socialism from Afar.* Ithaca: Cornell University Press.

Zhao, P., & Li, S. (2016). Restraining transport inequality in growing cities: Can spatial planning play a role? *International Journal of Sustainable Transportation, 10,* 947–995.

CHAPTER 12

Towards a Feminist Transport and Mobility Future: From One to Many Tracks

Tanja Joelsson and Christina Lindkvist Scholten

The 'mobility turn' in the social sciences, introduced by Sheller and Urry (2006), has played a significant role in opening up new perspectives for understanding mobility (Hannam et al. 2006). By broadening the understanding of flows—whether referring to transport from a conventional perspective, or as virtual travels or information transmission between remote places—the sociological understanding and ontology has contributed to an acknowledgement of a qualitative understanding of the meaning of transport and mobilities. As Sheller (2012) has pointed out, transport research is central to the development of the 'new mobilities' paradigm in the social and cultural sciences, but it 'exists in an expanded perspective using trans-disciplinary methodologies' (Sheller 2012: 290).

T. Joelsson (✉)
Department of Education, Uppsala University, Uppsala, Sweden
e-mail: tanja.joelsson@edu.uu.se

C. L. Scholten
Malmö University, Malmö, Sweden

K2 – Swedish Knowledge Centre for Public Transport, Lund, Sweden
e-mail: christina.scholten@mau.se; christina.scholten@k2centrum.se

© The Author(s) 2019   271
C. L. Scholten, T. Joelsson (eds.), *Integrating Gender into Transport Planning*, https://doi.org/10.1007/978-3-030-05042-9_12

In parallel, feminist research and other critical research traditions have pinpointed the necessity of taking gendered power relations seriously in social analyses of spaces and flows. Since the 1990s, new perspectives on transport have developed based on critical and feminist epistemologies. Women's and men's different modes of transport have been acknowledged, and analyses of the purposes of the trips made by women and men have made visible the gendered conditions of everyday life (Hanson and Pratt 1995; Law 1999; Dobbs 2007; Schwanen et al. 2008; Uteng and Cresswell 2008; Hanson 2010; Balkmar 2012; Scholten et al. 2012; Joelsson 2013). The travel choices and needs of different target groups, for example, the elderly (Levin 2008; Levin and Faith-Ell 2011; Berg et al. 2014) or young people (Brown et al. 2008; Barker et al. 2009; Fyhri et al. 2011; Salon and Gulyani 2010), have also been addressed in this research. This collection builds on these critical foundations and questions the dominant strands of politics in the field of transport planning research by exploring the normative prerequisites regarding transport and mobility.

Too often, research rooted in different epistemologies is solely concerned with sharing and developing knowledge within its own field. In this collection, the ambition has been to bring together different perspectives, despite the differences and difficulties related to different processes of knowledge production. We have a strong belief that it is necessary to find productive and constructive forms of sharing and understanding, and to make use of a multitude of methodologies to achieve changes towards more socially sustainable and equal transport solutions. The contributions in this collection range from cultural analysis and qualitative research epistemologies—in which cultural policy and discourse analysis are important methods to reveal hidden power dimensions in the creation of policy and decision-making—to quantitative analyses of statistical data sets. What the chapters have in common is an engagement with gender, transport and mobility planning. Throughout the volume, the reader can learn how women and men meet their transport needs differently, and can discover the fundamental importance of the political in defining, labelling and identifying the parameters when collecting statistical data (Smidfelt Rosqvist and Sánchez de Madariaga & Zucchini, in this collection). Data sets and statistical analyses provide an excellent understanding of major trends, such as commuting and its implications for family life (Sandow, in this collection), but a lack of numerical data is also a frequent problem, as discussed by Joelsson (in this collection). The knowledge produced evidently affects planning processes, and careful attention needs to be paid to collecting more context-sensitive data, as well as the critical assessment of existing transport and travel data. It has also become clear in this collection that not only gender but also other socially organizing principles affect and are affected by transport policy (Greed, Levy, Henriksson, Balkmar, Joelsson and Rømer Christensen, in this collection). It is therefore essential to address parallel and intersecting power orders, although at times the need to focus on gender may be politically strategic (Spivak and Guha 1988).

Many of the authors in this collection show that existing knowledge on gender seems to have trouble entering, and affecting, decision-making arenas. But it is not only the knowledge produced around transport practices that needs to relate to more complex realities. In order to promote democracy, planning processes and practices need to become more inclusive of citizen participation. It is vital to be able to present and communicate information and guiding documents related to proposed changes and adjustments to the transport system in a format that is appropriate for the citizens concerned in order to foster a transparent process.

PRESSING CONCERNS FOR POLICY AND PLANNING IN THE TRANSPORT FIELD

Critical epistemologies have furthermore raised issues of justice and social exclusion in transportation (Currie et al. 2009; Jones and Lucas 2012; Martens et al. 2012; Martens 2012, 2016; Lucas 2004, 2006, 2012). The perspective of social exclusion in transport has developed from its early stages of identifying the lack of affordable and accessible transport for the individual, into more complex analyses of land use and transport planning practices, and citizens' participation in transport planning decision-making processes (Lucas 2013). The poorer segments of society face challenges due to living in neighbourhoods where public transport is altogether lacking or only operates during inconvenient hours. Research on transport exclusion (Lucas 2004; Hine and Mitchell 2001; Social Exclusion Unit 2003; Levy, in this collection) has also shown that poor people commute further to their workplaces, which makes their travelling more complex (see Levy in this collection; Lucas 2012). Children's unaccompanied mobility is steadily decreasing, and children in better-off families are chauffeured or kept at home as a precaution against imagined dangers, resulting in children's 'retreat from the street' (Valentine 2004). Alongside the identification of vulnerable social groups in society, research on transport-related social exclusion has also identified core dimensions related to its impact on various social groups in relation to the following: personal, geographical, spatial, temporal, economic, social, cultural and political dimensions (Lucas 2013). Research on social exclusion has accelerated due to the British government's understanding of the importance of transport for citizens and the unfair distribution of transport services. In the late 1990s, the British government appointed a commission to investigate national conditions relating to transport poverty and social exclusion. This resulted in the Social Exclusion Unit Report of 2003 (Lucas 2012). The commission's final report identified a number of aspects relating to poor access to transport and the consequent implications regarding life opportunities. This commission has done pioneering work, showing an in-depth understanding of the implications of having access to transport or not, the importance of the design and structure of the transport system, and what the layout of connectivity has for different user groups regarding everyday livelihood opportunities.

It is in the tensions and public arguments revolving around transport planning and investment in transport infrastructure that questions of justice and fairness in relation to power relations between different socioeconomic groups and corporate interests can be discussed and transformed. In most countries, transport is considered to be the backbone of economic development and growth. The neoliberal discourse on growth, advocating market-oriented planning and decision-making, does not support debate regarding what we would define as necessary questions about which interests are chosen and at whose expense when priorities are set. When infrastructure investments in the road network, high-speed train systems or new airports are made—with references to the necessity for growth, the creation of attractive places or even the promotion of sustainability—two distinct perspectives are visible: a neoclassical planning agenda and a neoliberal agenda (Kębłowski and Bassens 2018). Neither of these pays attention to the outcomes in terms of justice (Martens 2016) or the need for advocacy for marginalized groups in the transport system (Fainstein 2014). Kębłowski and Bassens (2018) pinpoint the need to scrutinize the politics that contribute to the problem of social inequality and an unfair transport system:

> Political economy approaches thereby demonstrate how capitalist relations underpin transport policy. Critical scholars have exposed how the deeply contradictive marriage between emission-cutting sustainable ecology and growth-based economics has put forward a combination of behavioral

stimuli, technological innovations, and market- based instruments that inevitably engender a social lock-in. (Gössling and Cohen 2014, cited in Kębłowski and Bassens 2018: 422)

In line with this, we argue that transport politics needs attention and must be challenged by feminist arguments that highlight power relations and the redistribution of resources (Fraser 2005) in the context of uneven societal conditions.

Moreover, the growing need for more sustainable modes of transport due to climate change must also be scrutinized. Climate change has been on the global political agenda for the last five decades (WHO Fact sheet 2017). Moreover, metropolitan and urban regions are developing into hazardous living environments (SOER 2015). Despite all of this, the political and practical changes towards more sustainable transport solutions are being actualized too slowly (Hull 2008). Hanson (2010) claims that in order to really develop more sustainable models for transport, the following are needed: broader perspectives on transport and mobility, and a questioning of normative assumptions regarding technology and gender, mobility and gender, transport and gender, and planning and gender. As she puts it:

In travel for all purposes, not just work, women also use the car less (Polk 2004; Vance and Iovanna 2007) and drive fewer miles than men do (Rosenbloom 2006). As well as holding cross-nationally, these gender differences tend to remain when socio-demographic variables like education, income, and marital status are held constant. (Hanson 2010: 12)

The Paris Agreement and UN global goals have created a situation wherein transport politics can be informed by research from within the social sciences and humanities on how transport, transport planning and mobility could foster a more socially sustainable and environmentally friendly development. Many researchers in the field of 'new mobilities' have not only highlighted how virtually all current transport systems in the world are founded on oil-dependence, but they also have engaged in a more fundamental critique of the 'system of automobility' (Böhm et al. 2006). Sheller (2012) argues that we need a 'twin transition', not only to more sustainable mobility but also to mobility that is more just and equal. She finds that these processes are in fact interdependent and:

that a full transition in the currently dominant automobility system will only take place when we simultaneously address the issues of social inequality that underpin the un-sustainability of the current system, and begin to promote mobility justice as integral to sustainability. (Sheller 2012: 289)

The contributions in this collection support Sheller's advocacy for a twin transition, as the texts help to raise important questions on 'democracy, diversity and equity'—to paraphrase the urban planning theorist Susan Fainstein (2014)—in conjunction with issues of ecological sustainability. It is equally important, however, to recognize that the 'sustainability' discourse is indeed contradictory (Campbell 1996; Vallance et al. 2011). The conventional way of addressing transport politics and planning has, until quite recently and with some early exceptions, passed under the radar of more acute investigations based on acknowledging less powerful users of the transport system. Some of the contributors to this volume illustrate very clearly how reconceptualizations reveal new understandings of men's and women's mobility, while posing challenges for work on equality and sustainability (Greed, in this collection). As Rømer Christensen (in this collection, p. 264) puts it in relation to her investigation into men's and women's mobility in China:

> If, for example, women's more sustainable mobility is a reaction to the restrictions they experience and inequality of access, then we are confronted with a delicate paradox of equality and sustainability: enhancing gender equality within the present transport paradigm, in which cars are prioritized, would imply more women drivers, which would increase women's carbon footprint. Another strategy would be to urge men to change their mobility practices and to use women's mobility as a model, a suggestion which would be very difficult to implement. The paradox involved invites both the development of new research questions and the provision of a more robust knowledge base in various fields. This paradox might also be resolved through the introduction of new transformative modes and practices of transport. Here, new technologies such as bike-sharing might become part of the solution if they were combined with public interventions and strategies of social justice. The current situation in China, as in many other parts of the world, is that existing transport systems, with cars given priority, the inadequate provision of public transport and various new modes of private cycling, tend to leave women with more limited opportunities and choices than men. At the end of the day, this might prove to be detrimental to society, market potential and the quality of life for all.

The recent trend towards 'smart mobilities', to which Rømer Christensen alludes, may prove to be one example where potential shifts towards more equitable and sustainable mobilities can take place (Geoffron 2017; Song et al. 2017). It may also highlight competing and contradictory ideologies and discourses, focusing on cosmetic changes and leaving more structural aspects aside (Kitchin 2015; Araya 2015; cf. Joelsson forthcoming). What we hope has become clear so far is that the transition towards more sustainable and gender-equal transport systems is primarily, albeit not exclusively, a structural question (Greed, in this collection). By stressing the political nature of transport policy and planning, we are countering individualistic paradigms under which the responsibility for sustainability and equality is placed solely on the shoulders of individuals (Sheller and Urry 2016).

Given that this collection deals specifically with gender and transport planning, many related—but certainly no less significant—aspects have been shifted into the background. For instance, men and women encounter different public spaces while using the transport system. Women are harassed and violated by men—even raped and killed—or go in fear of being victimized while using public transport (Koskela 1999; Loukaitou-Sideri 2014; cf. Listerborn 2002). This is also true for non-white people, transgendered people and people identifying with non-normative sexualities, who may experience similar vulnerability when moving around in public space. For some, the private car may offer a safer space for movement than public transport, walking or cycling. Moving beyond (gender) stereotypes is of utmost importance, paying attention to the intersectional character of people's everyday lives. We hope to see more research in this area, as we recognize the void of knowledge about how transgendered people and people identifying with non-normative sexualities experience not only public space, but also public transport vis-á-vis other modes of transport in different environments. The growing body of research on accessibility and disability is another field and theme with which this collection has not directly engaged to any substantial extent (Currie and Stanley 2007; Wretstrand 2007), although the conceptual discussions on social inclusion and social sustainability are useful in this respect.

From One to Many Tracks

In this collection, we have focused particularly on planning and policies on mobility and transport—as these are transformed into the politics of infrastructure, the built environment and the design of transport systems—and how they have a fundamental impact on everyday mobilities and how people move, mainly from a Global North perspective. We have argued that planning exercises and policy work are highly political acts, which contribute to shaping both environments and how subjects move. There is hence an increasingly urgent need to develop political visions and ideas based on experiences and requirements that are contextualized and aware of the difficulties that women and men, and girls and boys, face when attempting to reach their everyday destinations. Politicians and planners need to recognize that the mainstream transport planning paradigm must evolve to create a fairer, more socially just and gender-equal transport system. It is crucial to grasp the complexity and context, encompassing a multitude of analyses with sometimes contradictory epistemologies, if we are to transition to a more just and sustainable transport society. Recognizing the potential for change in the shaping processes is as vital for the discussion on the impact of transport planning as is acknowledging the different conditions faced by different people depending on gender, age, race, ethnicity, class, sexuality, geographical location and so on. This is also what our title alludes to: the ambition to move from one track to many tracks.

The contributions made by this collection are threefold. Firstly, the various authors integrate gender research and transport planning, thus filling a research gap. Secondly, they combine quantitative and qualitative gender research perspectives and methods. And thirdly, the texts highlight the need to acknowledge the political within transport planning and transport practice. However, much remains to be done if we are to reach global objectives on equity and equality in relation to transport as a tool for improvement in social justice and combating climate change (see Levy, in this collection; Lucas 2011).

Having said all this, we wish to end this final chapter by listing some concrete suggestions for policy-makers and planners when it comes to integrating gender into transport planning in order to promote a more fair, just and equal transport system. The aspects discussed below are listed separately for the purpose of clarity, but are very much intertwined and dependent on one another.

1. Better, more nuanced and more detailed knowledge about transport issues can be gained through more finely tuned research. This entails engaging with critical perspectives, combining quantitative and qualitative research methods, in transdisciplinary and interdisciplinary research collaborations. Stakeholders from both within and out-

side academia are needed in this endeavour. One aim is to produce research demonstrating the necessity of understanding how gender and other organizing principles affect and are affected by the transport system. This will in turn assist in deconstructing the rationalist paradigm in which the natural sciences take precedence in knowledge about the world.

2. Existing knowledge about gender equality and gender research on transport and mobility need to be seen as relevant in order to become integrated into the knowledge base used in decision-making processes. One way of working towards this end is to integrate gender into undergraduate and graduate programmes in transport planning. In this way, the competence of planners can be improved. The competences of policy-makers must also be developed.

3. Clearer operationalization is needed of social categories such as gender, which may foster the use of social categories in planning and policy-making practice.

Finally, we hope that you have found this collection stimulating, thought-provoking and perhaps challenging, and that its findings and recommendations will help to encourage a new generation of transport planners to be truly inclusive as they map out our future mobility.

REFERENCES

Araya, D. (2015). *Smart cities as democratic ecologies*. London: Palgrave Macmillan Limited.

Balkmar, D. (2012). *On men and cars: An ethnographic study of gendered, risky and dangerous relations*. PhD dissertation, Linköping University, Sweden.

Barker, J., Kraftl, P., Horton, J., & Tucker, F. (2009). The road less travelled: New directions in children's and young people's mobility. *Mobilities, 4*(1), 1–10.

Berg, J., Levin, L., Abramsson, M., & Hagberg, J.-E. (2014). Mobility in the transition to retirement: The intertwining of transportation and everyday projects. *Journal of Transport Geography, 38*, 48–54.

Böhm, S., Jones, C., Land, C., & Paterson, M. (2006). *Against automobility*. Malden/Oxford/Carlton: Blackwell Publishing.

Brown, B., Mackett, R., Gong, Y., Kitazawa, K., & Paskins, J. (2008). Gender differences in children's pathways to independent mobility. *Children's Geographies, 6*(4), 385–401.

Campbell, S. (1996). Green cities, growing cities, just cities? Urban planning and the contradictions of sustainable development. *Journal of the American Planning Association, 62*(3), 296–312.

Currie, G., & Stanley, J. (2007). *No way to go: Transport and social disadvantage in Australian communities*. Melbourne: Monash University ePress.

Currie, G., Richardson, T., Smyth, P., Vella-Brodrick, D., Hine, J., Lucas, K., Stanley, J., Morris, J., Kinnear, R., & Stanley, J. (2009). Investigating links between transport disadvantage, social exclusion and well-being in Melbourne—Preliminary results. *Transport Policy, 16*(3), 97–105.

Dobbs, L. (2007). Stuck in the slow lane: Reconceptualizing the links between gender, transport and employment. *Gender, Work and Organization, 14*(2), 85–108.

Fainstein, S. S. (2014). The just city. *International Journal of Urban Sciences, 18*(1), 1–18.

Fraser, N. (2005). Reframing justice: In a globalizing world. *New Left Review, 36*, 69–88.

Fyhri, A., Hjorthol, R., Mackett, R. L., Fotel, T. N., & Kyttä, M. (2011). Children's active travel and independent mobility in four countries: Development, social contributing trends and measures. *Transport Policy, 18*(5), 703–710.

Geoffron, P. (2017). Smart cities and smart mobilities. In D. Attias (Ed.), *The automobile revolution* (pp. 87–98). Cham: Springer International Publishing.

Gössling, S., & Cohen, S. (2014). Why sustainable transport policies will fail: EU climate policy in the light of transport taboos. *Journal of Transport Geography, 39*, 197–207.

Hannam, K., Sheller, M., & Urry, J. (2006). Editorial: Mobilities, immobilities and moorings. *Mobilities, 1*(1), 1–22.

Hanson, S. (2010). Gender and mobility: New approaches for informing sustainability. *Gender, Place & Culture, 17*(1), 5–23.

Hanson, S., & Pratt, G. J. (1995). *Gender, work, and space*. London: Routledge.

Hine, J., & Mitchell, F. (2001). Better for everyone? Travel experiences and transport exclusion. *Urban Studies, 38*(2), 319–332.

Hull, A. (2008). Policy integration: What will it take to achieve more sustainable transport solutions in cities? *Transport Policy, 15*(2), 94–103.

Joelsson, T. (2013). *Space and sensibility: Young men's risk-taking with motor vehicles*. PhD dissertation, Linköping University, Sweden.

Joelsson, T. (forthcoming). Smart cities, smart mobilities and children. In T. P. Uteng, L. Levin, & H. R. Christensen (Eds.), *Gendering smart mobilities*. London: Routledge.

Jones, P., & Lucas, K. (2012). The social consequences of transport decision-making: Clarifying concepts, synthesizing knowledge and assessing implications. *Journal of Transport Geography, 21*, 4–16.

Kębłowski, W., & Bassens, D. (2018). "All transport problems are essentially mathematical": The uneven resonance of academic transport and mobility knowledge in Brussels. *Urban Geography, 39*(3), 413–437.

Kitchin, R. (2015). Making sense of smart cities: Addressing present shortcomings. *Cambridge Journal of Regions, Economics and Society, 8*, 131–136.

Koskela, H. (1999). 'Gendered exclusions': Women's fear of violence and changing relations to space. *Geografiska Annaler, 81B*(2), 111–124.

Law, R. (1999). Beyond 'women and transport': Towards new geographies of gender and daily mobility. *Progress in Human Geography, 23*(4), 567–588.

Levin, L. (2008). Äldre kvinnor: osynliga i statistiken men närvarande i trafiken [Elderly women: Invisible in statistics but present in the traffic]. In M. Brusman, T. Friberg, & J. Summerton (Eds.), *Resande, planering, makt* [Travel, planning, power]. Lund: Arkiv Förlag.

Levin, L., & Faith-Ell, C. (2011). *Women and men in public consultations of road-building projects*. Transportation Research Board Conference Proceedings.

Listerborn, C. (2002). *Trygg stad. Diskurser om rädsla i forskning, policyutveckling och lokal praktik* [Safe city: Discourses on women's fear in research, policy development and local practices]. PhD dissertation, Chalmers University of Technology, Sweden.

Loukaitou-Sideri, A. (2014). Fear and safety in transit environments from the women's perspective. *Security Journal, 27*(2), 242–256.

Lucas, K. (2004). *Running on empty: Transport, social exclusion and environmental justice*. Bristol: Policy Press.

Lucas, K. (2006). Providing transport for social inclusion within a framework for environmental justice in the UK. *Transportation Research Part A: Policy and Practice, 40*(10), 801–809.

Lucas, K. (2011). Making the connections between transport disadvantage and the social exclusion of low income populations in the Tshwane region of South Africa. *Journal of Transport Geography, 19*(6), 1320–1334.

Lucas, K. (2012). Transport and social exclusion: Where are we now? *Transport Policy, 20*, 105–113.

Lucas, K. (2013). Transport and social inclusion. In J.-P. Rodrigue, T. Notteboom, & J. Shaw (Eds.), *The SAGE handbook of transport studies*. London/Thousand Oaks/New Delhi/Singapore: Sage.

Martens, K. (2012). Justice in transport as justice in accessibility: Applying Walzer's 'spheres of justice' to the transport sector. *Transportation, 39*(6), 1035–1053.

Martens, K. (2016). *Transport justice: Designing fair transportation systems*. New York: Routledge.

Martens, K., Golub, A., & Robinson, G. (2012). A justice-theoretic approach to the distribution of transportation benefits: Implications for transportation planning practice in the United States. *Transportation Research Part A: Policy and Practice, 46*(4), 684–695.

Polk, M. (2004). The influence of gender on daily car use and on willingness to reduce car use in Sweden. *Journal of Transport Geography, 12*, 185–195.

Rosenbloom, S. (2006). Understanding women's and men's travel patterns: The research challenge. In *Research on women's issues in transportation: Volume 1 conference overview and plenary papers, conference proceedings* (Vol. 35, pp. 7–28). Washington, DC: National Research Council.

Salon, D., & Gulyani, S. (2010). Mobility, poverty, and gender: Travel 'choices' of slum residents in Nairobi, Kenya. *Transport Reviews, 30*(5), 641–657.

Scholten, C., Friberg, T., & Sandén, A. (2012). Re-reading time-geography from a gender perspective: Examples from gendered mobility. *Tijdschrift voor Economische en Sociale Geografie, 103*(5), 584–600.

Schwanen, T., Kwan, M.-P., & Ren, F. (2008). How fixed is fixed? Gendered rigidity of space–time constraints and geographies of everyday activities. *Geoforum, 39*, 2109–2121.

Social Exclusion Unit. 2003. *Making the connections: Final report on transport and social exclusion, 2003.* http://webarchive.nationalarchives.gov.uk/ and http://www.cabinetoffice.gov.uk/media/cabinetoffice/social_exclusion_task_force/assets/publications_1997_to_2006/making_transport_2003.pdf, 67.

Sheller, M. (2012). Sustainable mobility and mobility justice: Towards a twin transition. In J. Urry & M. Grieco (Eds.), *Mobilities: New perspectives on transport and society*. Farnham/Burlington: Taylor and Francis.

Sheller, M., & Urry, J. (2006). The new mobilities paradigm *Environment and Planning A, 38*(2), 207–226.

Sheller, M., & Urry, J. (2016). Mobilizing the new mobilities paradigm. *Applied Mobilities, 1*(1), 10–25.

SOER. (2015). European environment agency. Downloaded from: https://www.eea.europa.eu/soer-2015/europe/urban-systems. 20 Jan 2018.

Song, H., Srinivasan, R., Sookoor, T., & Jeschke, S. (2017). *Smart cities: Foundations and principles*. New York: Wiley.

Spivak, G. C., & Guha, R. (Eds.). (1988). *Selected subaltern studies*. Oxford: Oxford University Press.

Uteng, T. P., & Cresswell, T. E. (2008). *Gendered mobilities*. Aldershot: Ashgate.

Valentine, G. (2004). *Public space and the culture of childhood*. Aldershot: Ashgate.

Vallance, S., Perkins, H. C., & Dixon, J. E. (2011). What is social sustainability? A clarification of concepts. *Geoforum, 42*(3), 342–348.

Vance, C., & Iovanna, R. (2007). Gender and the automobile: Analysis of non-work service trips. *Transportation Research Record, 2013*, 54–61.

WHO Fact Sheet. (2017). *Climate change and health.* Downloaded from: http://www.who.int/news-room/fact-sheets/detail/climate-change-and-health. 20 Jan 2018.

Wretstrand, A. (2007). Comfort and safety as perceived by wheelchair-seated bus passengers. *Transportation Planning and Technology, 30*(2–3), 205–224.

Index[1]

[1] Note: Page numbers followed by 'n' refer to notes.

© The Author(s) 2019

C. L. Scholten, T. Joelsson (eds.), *Integrating Gender into Transport
Planning*, https://doi.org/10.1007/978-3-030-05042-9

Printed by Printforce, the Netherlands